FUNDAMENTAL BIOMEDICAL TECHNOLOGIES

*Series Editor*:
Mauro Ferrari, Ph.D. Houston, TX



For further volumes:
http://www.springer.com/series/7045

Gabriel A. Silva • Vladimir Parpura
Editors

# Nanotechnology for Biology and Medicine

## At the Building Block Level

Springer

*Editors*
Gabriel A. Silva
Departments of Bioengineering
and Ophthalmology
and Neurosciences Program
University of California
San Diego, CA, USA
gsilva@ucsd.edu

Vladimir Parpura
Department of Neurobiology
Atomic Force Microscopy
& Nanotechnology Laboratories
Center for Glial Biology in Medicine
Civitan International Research Center
Evelyn F. McKnight Brain Institute
University of Alabama
Birmingham, AL, USA
vlad@uab.edu

ISSN 1559-7083
ISBN 978-1-4614-3023-0 ISBN 978-0-387-31296-5 (eBook)
DOI 10.1007/978-0-387-31296-5
Springer New York Dordrecht Heidelberg London



Printed on acid-free paper

Springer is part of Springer Science+Business Media (www.springer.com)

*To Monica, Katie, and Jason*

*Gabriel A. Silva*

*To Vedrana, Vuga, and Ivan*

*Vladimir Parpura*

# Preface

Only within the last few years has much of the vision of Richard Feynman's now famous lecture on December 29th, 1959 at the American Physical Society meeting held at the California Institute of Technology, "There's plenty of room at the bottom," been realized. To a significant degree, this has been due to the relatively recent advances in chemistry, physics, materials science, and engineering that needed to precede the development of the kinds of technologies and applications Feynman eluded to in his talk. Designing, modeling, understanding, and ultimately building nanotechnologies is not an easy pursuit. Roughly speaking, nanotechnologies are engineered technologies and devices made up of materials and components over which some spatial aspect of the technology has been purposely designed and engineered at a nanoscale, typically regarded as between 1 and 100 nm or so. Importantly, the technology must exhibit some property or behavior or ability to interact with its environment that is unique and novel to the engineered device that is not a property of the constituent building block materials or elements. It is these properties from which the potential of nanotechnology stems. The scale at which such technologies are able to interact with their environment in order to produce unique and novel macroscopic effects is unprecedented.

Arguably one of the most important areas of application for nanotechnology is in biology and medicine. This is largely due to the vast complexity, both in structure and function, of biological systems, which makes developing technologies that can interface, sense, and respond to such systems difficult to design and engineer. In many ways, we know so little still about such systems that it is unclear how best to develop approaches and technologies to interact with them. Even in cases where the molecular and cellular details are understood for the most part, the correct materials, approaches, and methods to optimize such interactions are never clear cut. The potential of nanotechnology to accomplish such interactions results from the scale at which such technologies are designed relative to the fundamental functional scales at which cells operate. Most of the physical and chemical interactions between nanotechnologies and cells occur at the fundamental building molecular scale block level from which cells are made up. This ability translates into controlled interactions at a scale at which basic cell processes and functions are taking place, which ultimately translate up scales of organization to affect processes and functions at the tissue, organ, organ system, and ultimately organism levels. Thus, in theory, manipulations of materials and the design and engineering of devices at near atomic scales, in part the realization of Feynman's seminal lecture, should ultimately be able to accomplish these ideals and result in technologies that truly could, by design, interact with biological systems at fundamental building block scales. The result of such work will be spectacular, producing novel ways of measuring and studying biological systems in order to gain insights and understanding not possible with current technologies, and the development of technologies and methods for practical applications and uses, such as completely new ways to diagnose and treat diseases.

In practice though, much work remains to be done for such research to translate into viable experimental and clinical applications. As mentioned above, going from theory to practice is difficult. There are very few areas of modern science that are as truly interdisciplinary as nanotechnology and nanoengineering. The design, development, and eventual use of such technologies require the participation and coordination of individuals with different types of training and expertise. No one person has enough PhDs to cover every aspect of a nanotechnology project from start to finish. Given the proliferation of research centers, institutes, in some cases even departments in nanotechnology and nanoengineering, many students are now pursuing graduate degrees in nanotechnology. Despite this, their expertise will inevitably focus on one or a few areas, usually in the design and synthesis aspects. This means that no matter how one is trained, nanotechnology demands the input and participation of different individuals. This results in an interesting scientific sociological phenomenon: Very different scientific cultures and languages need to work together in order to accomplish meaningful high impact work. While this is true of any interdisciplinary pursuit, it is especially true and challenging for nanotechnology because it crosses so many basic and applied fields, all of them essential to the final result. For applications to biology and medicine, this means that biologists and clinicians from different disciplines need to be able to communicate and work with physical scientists. While it seems like a straightforward and obvious statement, anyone who has tried it knows how difficult it really is. Different disciplines have different values and a sense of what details are important, read different journals and attend different conferences, and individuals have different levels of understanding of areas outside of their comfort zone of expertise. Nonetheless, while challenging, it is all very doable, and in recent years we have seen some amazing advances from such collaborations, in particular, in areas of biology and medicine. Still, much of the work to date in the literature is still at the "proof of concept" stages. Very few applications have reached maturity. A colleague recently pointed out with some exasperation to one of us how it seems like the number of review articles populating the literature far exceeds the actual number of truly high impact primary research papers. In other words, while the potential for nanotechnology to have a high impact on biological, including biomedical research, and clinical medicine is very significant, very little to date has actually achieved such impact. This is not a negative, however, but an opportunity. Much of the research coming over the horizon seems truly spectacular, and we anticipate that the next few years will live up to this.

For example, in our specific area of interest, which is the application of nanotechnology to neuroscience, these methods and technology offer tremendous opportunities for new ways to measure and observe brain function at cellular and subcellular, i.e., molecular, scales. This in turn should allow us to understand neural dynamics and physiology in completely new ways, ultimately informing us about how the brain works under normal conditions and how it fails in cases of disease. Many of these nanotechnology methods for studying brain function are complementary to other emerging methods in neuroscience, such as genetic methods, although little has been done toward integrating them so far. In parallel, other approaches are focused on diagnostic and therapeutic targets that make use of nanotechnology. Many neurological disorders are multidimensional and mechanistically complex, and current therapeutic standards detect changes simply too late in the disease process for therapies to have significant effects. Similarly, given the complexity of neurological disorders, nanotechnologies provide the opportunity to perform different necessary functions in specific sequences: for example, crossing the blood brain barrier, targeting a cell population, and then delivering a therapeutic payload.

The contributions in this book have been written by some of the pioneers in the development and application of nanotechnology to biology and medicine. They are intended to provide the reader with an appreciation for and an understanding of the biological (i.e., molecular, cellular, and physiological) key challenges in this pursuit, as well as an understanding of the approaches and strategies being developed by cutting edge research to address such challenges. It is a reflection of and testament to the amount of progress that has occurred in an incredibly short period of time comparative to the long history of science, but it is also meant to give a realistic picture of the open problems and challenges that need to be resolved. In the process, though, we hope that different chapters will motivate and capture the imagination of the reader. Much work remains.

La Jolla, CA — Gabriel A. Silva
Birmingham, AL — Vladimir Parpura

# Acknowledgments

We would like to thank Diane Yu for her assistance. We thank all the authors for their contributions. This book would not exist without them.

# Contents

## Part IV Nanomedicine: Nanotechnology for Diagnosis and Treatment

# Contributors

**Laura Ballerini** Life Science Department, B.R.A.I.N., University of Trieste, Trieste, Italy

**Stephane Campidelli** Laboratoire d'Electronique Moléculaire, CEA Saclay, Gif-sur-Yvette, France

**Michael Cho** Department of Bioengineering, University of Illinois at Chicago, Chicago, IL, USA

**Alberto Credi** Dipartimento di Chimica "G. Ciamician", Università di Bologna, Bologna, Italy

**Chiara Daraio** Division of Engineering and Applied Science, California Institute of Technology, Pasadena, CA, USA

**Tejal A. Desai** Department of Bioengineering and Therapeutic Sciences, University of California, San Francisco, CA, USA

**Luca Gambazzi** Laboratory of Neural Microcircuitry, Brain Mind Institute, École Polytechnique Fédérale de Lausanne, Lausanne, Switzerland

**Fabrizio Gelain** Center for Biomedical Engineering, Center for Bits & Atoms, Massachusetts Institute of Technology, Cambridge, MA, USA

**Michele Giugliano** Laboratory of Neural Microcircuitry, Brain Mind Institute, École Polytechnique Fédérale de Lausanne, Lausanne, Switzerland

Department of Biomedical Sciences, University of Antwerp, Belgium

**Akihiro Horii** Center for Biomedical Engineering, Center for Bits & Atoms, Massachusetts Institute of Technology, Cambridge, MA, USA

**Sungho Jin** Materials Science & Engineering Program, Mechanical and Aerospace Engineering Department, University of California, San Diego, La Jolla, CA, USA

**Ratnesh Lal** Departments of Mechanical and Aerospace, and Bio-engineering, University of California, San Diego, La Jolla, CA, USA

**James J. Norman** Warner Babcock Institute for Green Chemistry, LLC, Wilmington, MA, USA

**Vladimir Parpura** Department of Neurobiology, Atomic Force Microscopy & Nanotechnology Laboratories, Center for Glial Biology in Medicine, Civitan International Research Center, Evelyn F. McKnight Brain Institute, University of Alabama, Birmingham, AL, USA;
IKERBASQUE, Basque Foundation for Science, Bilbao, Spain;
School of Medicine, University of Split, Split, Croatia

**Maurizio Prato** Department of Pharmaceutical Sciences, University of Trieste, Trieste, Italy

**Arjan P. Quist** Richmond Chemical Corporation, Oak Brook, IL, USA

**Sujata Sundara Rajan** Department of Biomedical Engineering, Oregon Health & Sciences University, Portland, OR, USA

**Reno C. Reyes** Department of Neurobiology, Atomic Force Microscopy & Nanotechnology Laboratories, Center for Glial Biology in Medicine, Civitan International Research Center, Evelyn F. McKnight Brain Institute, University of Alabama, Birmingham, AL, USA;
Department of Neurology, University of California, San Francisco and Veterans Affairs Medical Center, San Francisco, CA, USA

**Gabriel A. Silva** Departments of Bioengineering and Ophthalmology and Neurosciences Program, University of California, San Diego, CA, USA

**Serena Silvi** Dipartimento di Chimica "G. Ciamician", Università di Bologna, Bologna, Italy

**Shan Sun** Department of Bioengineering, University of Illinois at Chicago, Chicago, IL, USA

**Igor Titushkin** Department of Bioengineering, University of Illinois at Chicago, Chicago, IL, USA

**Tania Q. Vu** Department of Biomedical Engineering, Oregon Health & Sciences University, Portland, OR, USA

**Hidenori Yokoi** Center for Biomedical Engineering, Center for Bits & Atoms, Massachusetts Institute of Technology, Cambridge, MA, USA

**Shuguang Zhang** Center for Biomedical Engineering, Center for Bits & Atoms, Massachusetts Institute of Technology, Cambridge, MA, USA

# Part I
# Nanoscale Processes in Cells

# Structure and Biology of the Cellular Environment: The Extracellular Matrix

**Igor Titushkin, Shan Sun, and Michael Cho**

## 1 Introduction

The extracellular matrix (ECM) represents a complex organization of macromolecules that surrounds the cell and comprises the substratum onto which the cell may be attached. The properties and functions of the ECM depend ultimately on its structure, molecular components, architecture, and dynamic modulation. Because the critical role of ECM involved in cell biology and physiology has long been recognized, the structure and biology of the ECM have been extensively studied (Yurchenco and Birk 1994; Ayad et al. 1998; Robert 2001). The diversity found in the structure and organization of the ECM appears to be tissue specific and regulates the properties and function of each tissue. The ECM was once believed to provide mainly the structural support and tensile strength of the tissue. It is now recognized that in addition to the role of mechanical stability, the ECM also provides substrate for cell adhesion and migration and regulates cell differentiation, metabolic activity, and cell–cell signaling, and therefore intimately involved in cellular and molecular response and behavior of the cell. For example, the ECM is found to harbor potent signaling cues, such as cytokines and growth factors (Engel 2004). Furthermore, the relationship between cell and ECM can be considered bidirectional (Cukierman et al. 2001; Geiger et al. 2001; Labat-Robert 2004; Yoon et al. 2005): while ECM is known to influence the cell behaviors and functions, the cell, in turn, modifies local environment and remodels the ECM. This ability of ECM to undergo remodeling plays an important role especially in developmental tissue (Zagris 2001; Trelstad 2004) and wound healing (Schaffer and Nanney 1996; Tuan et al. 1996; Cutroneo 2003). In the context of tissue

I. Titushkin • S. Sun • M. Cho (✉)
Department of Bioengineering, University of Illinois at Chicago,
851 S. Morgan St. (M/C 063), Chicago, IL 60607, USA
e-mail: mcho@uic.edu

G.A. Silva and V. Parpura (eds.), *Nanotechnology for Biology and Medicine: At the Building Block Level*, Fundamental Biomedical Technologies,
DOI 10.1007/978-0-387-31296-5_1, © Springer Science+Business Media, LLC 2012

engineering, elucidation of the ECM structure and biology also has a rather critical implication. Dynamic interactions between ECM and cell must be characterized and understood at the molecular level for desired manipulation of the ECM that lead to successful tissue-engineering applications. Congruent with the current research emphasis on nanotechnology, tissue-engineering methodologies will likely be refined and improved by replication of the complex features of natural ECM at the micro- and nanoscale.

# 2 Composition and Architecture of ECM

## *2.1 Major ECM Molecules*

The composition of the ECM may be classified into at least five major categories of components. Molecules found in the ECM include (1) structural proteins, such as collagen and elastin; (2) adhesion proteins, such as fibronectin and laminin; (3) soluble growth factors and cytokines; (4) proteoglycans; and (5) tissue-specific molecules. First, collagen, with at least known 21 subtypes in its family, is the major protein that comprises the ECM. Each type-I collagen molecule, for instance, contains three chains (called α chains) twisted around each other to form a triple helix, and lateral interactions of triple helices of collagens result in the formation of fibrils roughly 10–300 nm diameter and up to hundreds of micron length in mature tissues (Alberts et al. 2002). The triple helix gives the collagen a rigid structure that maintains the mechanical and structural integrity of tissues. There exist several different α chains that are found in various combinations in the collagen subtypes. Second, fibronectin, the next largest quantity of proteins found in the ECM, is one of the glycoproteins involved in cell adhesion to the ECM and cell migration. It exists as a dimer composed of two very large subunits joined by a pair of disulfide bonds near the carboxyl termini (Alberts et al. 2002). Each subunit is folded into a series of functionally distinct domains separated by regions of flexible polypeptide chains 60–70-nm long and 2–3-nm thick. Third, growth factors and cytokines are the examples of signaling proteins solubilized in the ECM that can stimulate cells to grow, migrate, and mediate cell–cell communications (Perris and Perissinotto 2000; Brownlee 2002; Kleinman et al. 2003). These generally small signaling molecules can influence the cell by binding to a receptor expressed on the cell surface and activate the receptor-mediated signaling transduction cascades. Fourth, macromolecules found in the ECM include proteoglycans and glycosaminoglycan (GAG). Proteoglycans consist of one or more GAG chains attached to a core protein. Unlike glycoproteins with one or more oligosaccharide chains covalently bound to amino acid side chains, proteoglycans contain much more carbohydrate by weight, mostly in the form of long, unbranched, GAG chains rather than short, branched, oligosaccharide chains found in glycoproteins (Alberts et al. 2002). Proteoglycans are thought to play a key role in chemical signaling between cells, and regulate the activities of secreted proteins, such as growth factors. Some of the plasma membrane proteoglycans can bind cells to the ECM and trigger the responses of cells to extracellular signals (Simons and Horowitz 2001; Yoneda and Couchman 2003). GAG is an unbranched, negatively charged, polysaccharide chain composed by repeating disaccharide units. The main groups of GAGs of physiological significance are hyaluronic acid (hyaluronan), chondroitin sulfate, dermatan sulfate, heparan sulfate, and keratan sulfate. Hyaluronan is unique among the GAGs because it does not contain any sulfate and is a component of non-covalently formed complexes with proteoglycans in the ECM. Hyaluronic acid polymers are very large (with molecular weights of 100–10,000 kDa) and can occupy a large volume. GAGs are located either in the ECM or on the surface of cells, where these molecules serve as coreceptors to help cells respond to secreted signal proteins (Raman et al. 2005). Due to high viscosity and low compressibility, the mechanical characteristics of GAG are ideal for excellent lubricators and shock absorbers and used as a lubricating fluid, for example, in the joints. Finally, examples of the tissue-specific components include aggregans in the cartilage tissue or minerals (e.g., hydroxyapatite) in the bone tissue.

## *2.2 ECM Architecture*

Similar architectural characteristics of ECM can be found in the major tissue types, such as nerve, muscle, epithelial, and connective tissues. Generally, the ECM is made of various protein fibers interwoven in a hydrated gel composed of proteoglycans and GAG chains (Fig. 1). Fibrillar collagen forms the major matrix, strengthens the scaffold, and also provides substratum for cell adhesion. GAGs fill a large volume and form highly hydrated gels in ECM. Adhesion proteins in the matrix and on the surface of cell membrane bind macromolecules and cells to build up ECM into an active and dynamic organization that can influence the cellular cytoskeleton and cell spreading. The diversity of ECM in different tissues arises from the relative amounts of the macromolecules mentioned above, tissue-specific components, and the way in which they are arranged. For example, connective tissue and epithelial tissue represent two extremes of contrasting spatial organization. In connective tissue, cells are sparsely distributed within the ECM. Direct attachments between cells are relatively rare, and the ECM is rich in fibrous polymers, especially collagens, which bear most of the mechanical stress the tissue is subjected to. In contrast, epithelial tissue has a scant ECM, consisting mainly of a thin mat called the basal lamina, which underlies the epithelium. Cells are tightly bound together into sheets called epithelia. The cells are attached to each other by cell–cell adhesions that bear most of the mechanical stresses. Strong intracellular protein filaments connect the cells either to each other or to the underlying basal lamina. Examples would include capillaries that consist of a single layer of endothelial cells attached to a basal

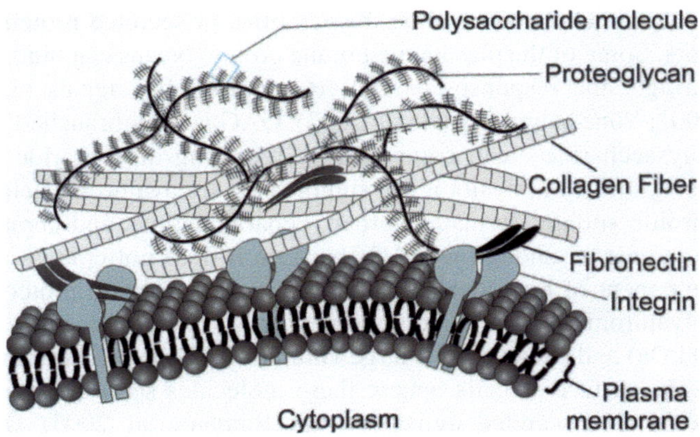


**Fig. 1** Schematic cartoon for ECM molecules and architectural arrangement

lamina and every smooth muscle that is attached to its neighboring muscle cells by a basal lamina.

## *2.3 ECM–Cell Communication Pathways*

Bidirectional communication between ECM and cell is likely mediated by a family of transmembrane adhesion proteins that are referred to as integrins, which act as receptors and bind to the ECM components, including collagen and fibronectin (Heino 2000; Farias et al. 2005). An integrin molecule has two non-covalently associated transmembrane glycoprotein subunits called $\alpha$ and $\beta$ chains with their globular heads projecting more than 20 nm from the lipid bilayer (Smith 1994). As this end of integrin binds the ECM with different specificity and relatively low affinity, the other end of integrin binds to actin cytoskeleton inside the cell. This dual binding capacity is required to relate the molecular signals in and out of the cell (i.e., bidirectional communication), thus providing active intracellular signaling pathways for communications between the cell and ECM. A unique feature of the integrin is found in its adhesive motif that was initially discovered by Pierschbacher et al. (1981). A 11-kDa fragment from fibronectin was identified to have the capacity to support cell adhesion. Subsequent studies (Pierschbacher and Ruoslahti 1984) showed that this fragment can be mimicked by a small peptide containing the Arg–Gly–Asp (RGD) sequence. In addition to fibronectin, this adhesive motif is found in other ECM proteins and blood adhesion proteins (Suzuki et al. 1985; Sadler et al. 1985). Not all, but most of the subtypes of integrins recognize this particular adhesive motif, and thus integrins can bind to the major ECM proteins (fibronectin, collagen, laminin, vitronectin, to name just a few) and mediate the receptor–ligand binding interactions.

# 3 Role of ECM in Tissue Development and Wound Healing

## *3.1 Expression of ECM in the Biological Process of Tissue Development*

In tissue development, the ECM molecules are known to direct cell attachment, movement, and localize inductive signals. The development of ECM reflects the acquisition of differentiated functions of the cells. Newly produced ECM in turn, via cell signaling, specifies cell fate and regulates the formation of tissues and morphogenesis of organs (Zagris 2001). Improper regulation of ECM can result in aberrant tissue development and diseases. Many ECM components appear only transiently during specific developmental or pathological events (Svetlana 2004). For example, laminins, which are heterotrimeric glycoproteins in the basement membrane, consist of multiple isoforms that vary in their chain composition and tissue distribution during embryogenesis (De Arcangelis and Labouesse 2000). More significantly, degradation and remodeling of the ECM are largely controlled by matrix metalloproteinases (or metalloproteases; MMPs) and their specific tissue inhibitors of metalloproteinases (TIMPs). Many MMPs [at least 20 subtypes identified (Jones et al. 2003)] are expressed widely during embryogenesis, but not in adult life. In adult life, MMPs are expressed in rapidly remodeling tissues, such as the term placenta, menstrual endometrium, and involuting mammary glands, and during wound healing. The expression of MMPs in organogenesis is intensively controlled by growth factors and cytokines (Feinberg et al. 2000). ECM molecules and their receptors are essential in development, because they regulate many aspects in tissue-specific development, such as cell growth and proliferation modulated by the growth factors. During embryonic development of multicellular organisms, integrins are the main family of cell-surface receptors that mediate cell–matrix interactions and signaling pathway. This adhesion function of ECM has often been mimicked in synthetic ECMs by fabricating the polymers to include the adhesive motif (e.g., RGD peptides) and presenting them to the cell. While fibronectin is required for the mesoderm development (De Arcangelis and Labouesse 2000), other ECM proteins, such as laminins and collagens, play a particularly crucial role in epithelial development, and the informational cues arising from ECM remodeling are transmitted via intracellular signaling to effect epithelial gene expression.

## *3.2 ECM of Wound Healing*

Similar to intricate dynamics of ECM remodeling during tissue development, another excellent example of ECM remodeling is found in wound healing. It is a complex process that involves different cell types and coordinated cellular activities and reveals a multifunctional ECM role in normal and injured tissue metabolism. For example, in skin wound repair, white cells, keratinocytes, fibroblasts, and

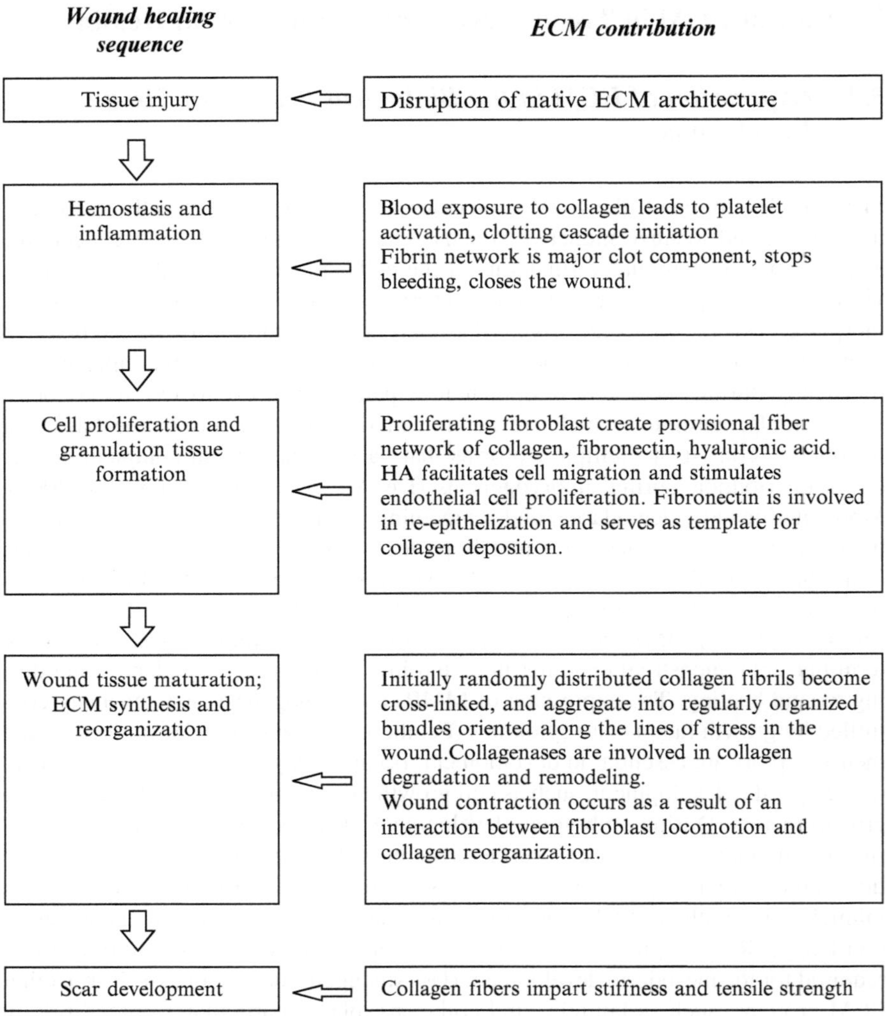


**Fig. 2** Block diagram of ECM remodeling in wound healing

endothelial cells all play a role in a well-established sequence of events that eventually lead to repair of the damage tissue by remodeling ECM. In a schematic that divides the wound healing into distinctive different stages (inflammation, proliferation, and remodeling; see Fig. 2), ECM remodeling in the wound begins as soon as a new ECM is laid down. Different cell types play a key role at each stage and are regulated by numerous growth factors and cytokines (Moore 2001). The progression of wound repair is frequently modeled in the skin, but parallel and temporally coordinated events occur in most tissues following injury.

Wound healing response is initiated at the moment of injury with the hemorrhage and formation of the blood clot. Immediately after that inflammation phase is triggered, which is characterized by prominent immune response involving neutrophils, macrophages, lymphocytes, and other immune elements. In the following proliferating phase, fibroblasts migrate into the area of wound to form the granulation tissue and to produce ECM, which is dynamically remodeled for effective wound closure and scar development. Simultaneously, wound contraction and re-epithelization take place. All these events are strictly regulated by wound microenvironment, including growth factors, cytokines, and newly created ECM components. The final product of the healing process is a scar – relatively avascular and acellular mass of collagen serving to restore tissue continuity, strength, and function. This pivotal role of ECM in wound healing sequences can be observed throughout all stages of the repair process.

Exposure of blood to collagen in the wound defects results in platelet activation and coagulation cascade initiation, while the final fibrin network clot serves to stop bleeding and to plug the wound site. A number of growth factors and cytokines are released from platelets, which play important role in recruitment of inflammatory cells, stimulation of fibroblasts, and evolution of granulation tissue. For example, platelet-derived growth factor (PDGF) and transforming growth factor (TGF)-β1 regulate many matrix proteins including collagen, proteoglycans, fibronectin, and matrix degrading proteases and their inhibitors. Latent TGF-β1, released from platelets and inflammatory cells, is activated by proteolytic and non-proteolytic mechanisms to influence wound healing from initial clot formation to the final stage of matrix deposition and remodeling (Wahl 1999). ECM also acts as a reservoir for growth factors required during healing. Fibroblast growth factors (FGF-1 and -2) are weakly soluble with a strong affinity for heparan sulfate. Pre-existing FGF-2 bound in the wounded ECM appears to stimulate angiogenesis and fibroblast function modulation (Mustoe et al. 1991). Hyaluronic acid, highly hydrated ECM component conferring viscosity to tissues, promotes early inflammation by enhancing leukocyte infiltration. It also moderates the inflammatory response as healing progresses towards granulation tissue formation and facilitates fibroblast migration. The neutrophils are activated during chemotaxis and produce elastase and collagenase to facilitate their migration. Once in the tissue, the inflammatory cells and fibroblasts stimulate the production of MMP-1, -2, -3, and -9 to degrade the damaged ECM in preparation for macrophage phagocytosis of the ECM debris (Mott and Werb 2004). After inflammation, the proliferative phase follows, which is characterized by granulation tissue formation. It consists of cellular elements (fibroblasts and inflammatory cells), along with new capillaries embedded in a loose ECM of collagen, fibronectin, and hyaluronan. Fibroblasts respond to cytokines/growth factors by proliferating and synthesizing provisional fiber network rich in fibronectin. It serves not only as a substratum for migration and in-growth of cells, but also as a template of collagen deposition by fibroblasts. There are also significant quantities of hyaluronic acid and large-molecular-weight proteoglycans present, which contribute to the gel-like consistency of the ECM and aid cellular infiltration. TGF-β1 further contributes to this fibrotic process by recruiting fibroblasts and stimulating their synthesis of collagens I, III, and V, proteoglycans, fibronectin, and other ECM components (Nissen et al. 1999). TGF-β1 concurrently inhibits proteases while enhancing protease inhibitors, favoring matrix accumulation. Re-epithelization at the wound surface and revascularization of the wound proceed in parallel with granulation tissue formation. For example, immature keratinocytes in skin produce MMPs and plasmin to dissociate from the basement membrane and facilitate their migration across the open wound bed in response to chemoattractants. The migration of epithelial cells occurs independently of proliferation and depends upon a number of factors, including growth factors, loss of contact with adjacent cells, and guidance by active contact with ECM. Adhesive ECM proteins, such as fibronectin or vitronectin, bind to keratinocytes and induce them to migrate over granulation tissue as part of wound re-epithelialization (O'Toole 2001). Subsequently, provisional matrix components, such as hyaluronic acid and fibronectin, are gradually replaced by collagen and proteoglycans.

The predominant collagen subtype in skin changes from type III (in the early healing wound) to type I in the mature wound. Collagen is constantly being degraded and resynthesized even in normal intact tissues. Following injury, the rate of collagen synthesis increases dramatically for approximately 3 weeks. The gradual gain in wound stiffness and tensile strength is due not only to continuing collagen deposition, but also to collagen remodeling. The initially, randomly distributed collagen fibers become cross-linked by enzyme lysyl oxidase and aggregated into regularly aligned fibrillar bundles, oriented along the line of stress of the healing wound. The net increase in wound collagen is determined by the balance of its synthesis and catabolism. The degradation of fibrillar collagen is driven by serine proteases and MMPs under the control of the cytokine network. Production of MMPs and TIMPs by fibroblasts is inducible and tightly regulated by cytokines, growth factors, hormones, and contact with ECM components.

Collagen is also closely involved in wound contraction – inward movement of the wound edges, which occurs as a result of an interaction between fibroblast locomotion and collagen reorganization. The contraction is mediated by attachment of collagen fibrils to cell-surface receptors. The resulting contraction force generated by cell motility brings the attached collagen fibrils closer together and eventually compacts them. A specialized subset of fibroblasts with muscle-like contractile features called myofibroblasts may be also involved in wound contraction (Desmouliere et al. 2005). As remodeling progresses, there is a gradual reduction in the cell number and vascularity of the reparative tissue which results in the formation of a relatively avascular and acellular collagen scar. Remodeling of the scar can continue for 1–2 years. The relative weakness of the scar compared to normal tissue is a consequence of the collagen fiber bundle orientation and abnormal molecular cross-linking. For example, because the fibers in normal tissue are relatively randomly organized rather than the fibers oriented parallel in scar, the maximum breaking strength of mature scar in skin is only 70% of the intact skin. In summary, the ECM dictates and directs wound healing to a great extent by inducing the cell stickiness, movement, proliferation, and differentiation. The growing body of knowledge of ECM metabolism in wound healing and how cells interact with it can help developing new treatment strategies to minimize scarring. It is fascinating to note that fetal wounds are healed without scar formation. Elucidation of the molecular mechanisms and physical and chemical cues that are responsible for scarless fetal wound healing would lead to ideal tissue repair, but currently remains unknown (Colwell et al. 2005). Identification of such mechanisms and cues would undoubtedly be useful for attempts to engineer a skin-equivalent tissue replacement construct. Indeed, one of the most successful tissue-engineering applications up to date has been the treatment of skin wound that incorporates keratinocytes and fibroblasts seeded in both natural and synthetic ECM. Advances in tissue engineering of tissue types other than skin are also rapidly progressing. Some recent developments of tissue engineering in creating ECM for tissue restoration and other biomedical applications are described in the following section. Finally, electrotherapy for wound healing has been applied over the past two decades. Although the molecular mechanisms that mediate electrically stimulated wound healing are yet to be fully elucidated, a recent review suggests that several cellular events are induced and coordinated by electrical stimulation to promote cell adhesion and migration that reorganizes ECM structures [see Cho (2002)) for review].

## 4 Role of ECM in Tissue Engineering

Based on the important and critical role of ECM to influence cellular and molecular responses, development of extracellular environment has become a key task for tissue engineering, regenerative medicine, and other biomedical and pharmaceutical applications. Successful tissue engineering requires appropriate integration of at least three critical components: cells; natural or synthetic scaffold that can serve as a temporary ECM; and chemical and physical elements, such as growth factors and topographical features that best mimic the extracellular environment in the natural ECM. Scaffold must be engineered in a way that is conducive to cell proliferation, differentiation, and eventual integration with the surrounding tissue. Designing a conducive extracellular environment requires better understanding of the role of biological, chemical, and physical stimuli that may influence the structural integrity and functionality of the engineered tissue. Use of biodegradable polymer scaffolds is generally preferred because the polymer scaffolds can serve as a template for tissue development and the scaffolds are resorbed, avoiding foreign body response. For example, collagen-based hydrogel has been extensively used as a scaffold for tissue engineering. There are several advantages for using the collagen-based scaffold. First, as described earlier, collagen is the principal component of in vivo ECM. Second, it provides a suitable microenvironment for cells, including large pores and high mechanical strength (Friess 1998; Yannas 1995). Third, cells can be directly incorporated into the collagen monomeric solution. Fourth, by varying the scaffold composition, the mesh size is controlled, and the optimal scaffold composition may be determined for engineering a specific tissue type. Fifth, collagen scaffold can be served a model to study the complex tissue development as the seeded cells interact

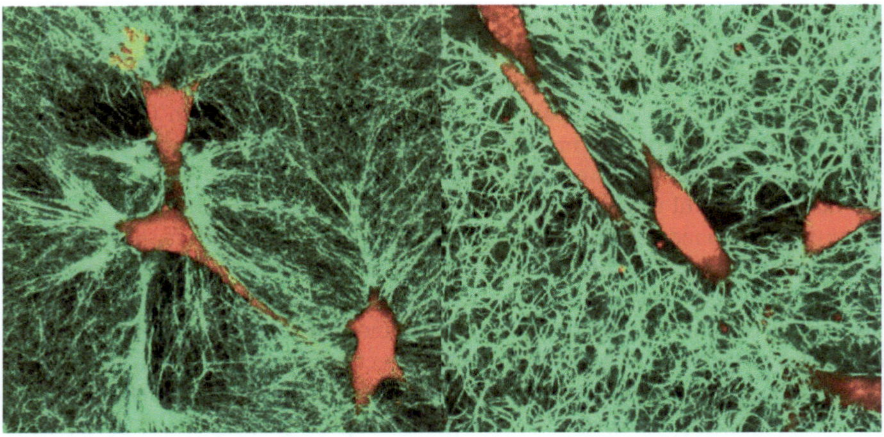

**Fig. 3** Collagen scaffold (1 mg/ml) structures and seeded cells were imaged using a Bio-Rad Radiance Model 2000 (femtosecond pulses at 80 MHz, 0.8 mW) laser scanning multiphoton microscope. Second harmonic generation signals from collagen fibers (no fluorophore required) were excited by a wavelength of 840 nm beam, and emitted signals were acquired with a band-pass filter of 390±70 nm. Cells were stained using the CellTracker orange CMTMR (Molecular Probe, OR, USA) and visualized with a band-pass filter of 560±70 nm. Spectrally separated but combined images of rat mesenchymal stem cells (*left panel*) and human fibroblasts (*right panel*) are shown

with and remodel the ECM. Finally, collagen-based hydrogels are relatively simple to construct and does not require expensive equipment. One major disadvantage of collagen-based scaffold could be the lack of strong mechanical strength essential for engineering, for example, bone tissue. Although type-I collagen constitutes ~90% of ECM in the bone tissue, collagen polymers alone are insufficient to provide the necessary mechanical strength of the bone tissue. Other materials, such as hydroxyapatite and calcium phosphate, have been added to the collagen scaffold to improve mechanical strength (McCarthy et al. 1996; Takoaka et al. 1998; LeGeros 2002; Alhadlaq et al. 2004). Using multiphoton microscopy, we were able to obtain simultaneously the images of collagen fibers and cells seeded in the 3D collagen scaffold, allowing us to determine collagen–cell binding. Collagen fibers were imaged without fluorescent probes, but instead relied on Second Harmonic Generation imaging technique (Zoumi et al. 2002; Zipfel et al. 2003), while cells were loaded with CellTrackers. Seeding rat mesenchymal stem cells (rMSCs) and human fibroblasts in the collagen scaffold, differential cell adhesion behaviors were investigated. As shown in Fig. 3, rMSCs appear to concentrate and reorient collagen fibers (left panel) that are consistent with the postulate of strong adhesion, whereas fibroblasts embedded the same collagen scaffold demonstrated less localized, but randomized collagen fibers (right panel). Moreover, the extent of collagen reorganization around the fibroblast appears not as dramatic as that found around the MSC.

While natural polymers, such as collagen and chitosan, closely mimic the native extracellular environment, synthetic polymers have been developed for tissue-engineering applications. Substantial advances have been achieved recently with synthetic biomaterials. These synthetic resorbable polymers provide much higher mechanical strength and are degraded by hydrolysis. Although synthetic polymers can be produced and used for tissue engineering, the FDA-approved biomaterials such as polylactic acid (PLA), polyglycolic acid (PGA), or mixture of these two polymers are by far the most commonly used biocompatible polymers. From the mechanical perspective, the Young's modulus of cortical bone is on the order of ~10 GPa, and the Young's modulus for PGA closely matches this value [~7 GPa (Yang et al. 2001)]. In addition to providing the mechanical strength, synthetic polymers can be manipulated to create and control the scaffold architecture. This has become an important task in tissue engineering because (1) spatial organization of cell seeding and cell growth depends on the scaffold architecture and (2) this architecture is now believed to regulate the development of specific tissue functions. Cima et al. (1991) demonstrated that tissue regeneration using synthetic materials depends on the porosity and pore size of the scaffold. While large surface area promotes cell attachment, large pores are desirable for providing sufficient nutrients and removing waste. Another aspect of the scaffold architecture is the degree of continuity of the pores. Molecule transport and cell migration have been shown to be prevented in the highly porous matrix of disconnected pores.

## *4.1 Microfabrication*

Many techniques have been developed over the years to physically control and regulate the topographical features of synthetic scaffold. One prominent approach that has become popular in the field of tissue engineering includes microfabrication. This technique relies on surface modification of biomaterials via physical or chemical methods that can lead to desired cellular responses (Wilkinson et al. 2002). Physical techniques, such as microcontact printing, casting, and embossing, have been known for a long time, and also successfully been downscaled to micro- or nanoworld (Curtis and Wilkinson 2001). Microcontact printing, shown in Fig. 4, uses a lithographically fabricated stamp (usually from polydimethylsiloxane) with the desired pattern, which can be inked with protein, polysaccharide, or other large molecule. On pressing the stamp onto a substrate preconditioned with a sticky layer, the molecule is transferred with the pattern of the stamp (Michel et al. 2002a, b). This simple and cost-effective approach results in chemical patterns with micron or submicron size features displaying binding sites for specific molecules. Examples of successful microfabrication applications include designing controlled drug delivery systems (Tao et al. 2003), altering surface topography to mimic in vivo conditions (Motlagh et al. 2003), microfluidic flow (Popat and Desai 2004), and 3D configuration of multilayer cell cultures (Tan and Desai 2004).

Microfabrication techniques offer a useful tool to control selective cell adhesion and spatial organization of cells in the 3D scaffold. Cell adhesion and motility is one of the important biological processes involved in cell growth, differentiation, inflammatory response, and wound healing, and is often desired for engineering tissue

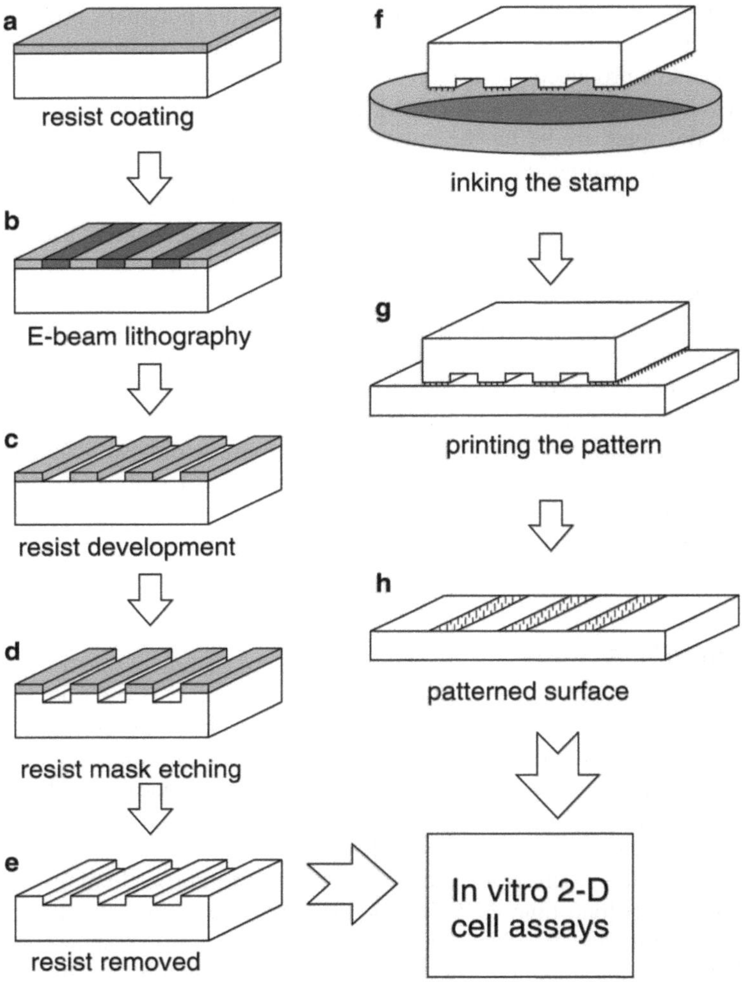


**Fig. 4** Schematic cartoon for micro- or nanopatterned substrates for cell engineering. First, the pattern is defined in a radiation-sensitive resist coated on the substrate (**a**) with electron beam or another lithographic technique (**b**). After resist development (**c**), it serves as a mask to transfer the pattern onto the substrate by dry or wet etching process (**d**). Resulting pattern can be used immediately for experiments with cells or as a stamp in contact micro- or nanoprinting (**f**). The stamp is inked with desired biomolecule or chemical agent and pressed onto a preconditioned adhesive surface (**e**). The ink solution is transferred to the surface with the pattern of the stamp (**h**). The uncoated areas can be further derivatized with another chemistry to control wettability, adhesion, biological specificity, and other material surface properties on nanoscale

constructs. In addition to modifying surface topography and chemistry to regulate cell adhesion, an alternate methodology of applying physical forces that can induce the similar cellular responses has been demonstrated. For example, because cellular behaviors have been shown to depend on the local electrical environment within tissues, we have incorporated the use of non-invasive electrical stimulus to characterize and control the human fibroblast adhesion and movement in the 3D collagen gel model (Sun et al. 2004). The cell movements in the 3D collagen gel were shown to depend on both electrical stimulus strength and collagen concentration. A small non-invasive electrical stimulus (0.1 V/cm) was found to be sufficient to induce 3D cell migration, while the collagen concentration of ~0.6 mg/ml appeared to represent the optimal scaffold network environment. However, the same electrical stimulus failed to induce 2D cell movement. This observation provides a clue that biophysically unconstrained 2D cell adhesion may represent the "exaggerated state" of cell adhesion, and that the sufficient and necessary cell adhesion found in the 3D collagen gel, which resembles in vivo cellular responses, is likely weaker than that found on 2D substrate. It is interesting to note that this induced human fibroblast movement in 3D collagen gel is both integrin- and $Ca^{2+}$ dependent (Sun and Cho 2004). Treatment of cells with anti-integrin antibodies prevents electrically induced cell movement. While the absence of extracellular $Ca^{2+}$ suppresses the 3D cell movement, inhibition of the cell-surface receptor-coupled phospholipase C (PLC) completely prevents 3D cell migration, suggesting molecular association among integrin, PLC, and intracellular $Ca^{2+}$. Elucidation of the electrocoupling molecular mechanisms involved 3D cell movement could lead to controlled and designed manipulation of 3D cell adhesion and migration, and may be used to complement the microfabrication techniques that have been successfully applied to tissue engineering.

## 4.2 *Nanofabrication*

Recent progress in nanotechnology provides a variety of methods for fabricating materials with nanostructure and nanochemistry similar to that naturally occurring in the cell surrounding. Both micro- or nanoengineering techniques are aimed to control physicochemical and biological properties of synthetic materials, which are thought to be essential for guiding complex cellular processes and directing cell behavior and functions (Curtis 2004; Lutolf and Hubbell 2005). These nanoscale features include, but not limited to distribution of receptor-binding ligands and surface-bound growth factors, surface nanotopography, presentation of domains susceptible for cell-triggered degradation, and remodeling. Molecular mechanisms found in the natural ECM that regulate cellular functions are being incorporated in fabrication of synthetic extracellular environment at nanoscale. For example, mounting data support the notion that cells respond to the precise nanoscale spatial organization of adhesion ligands (Irvine et al. 2002). This is perhaps related to the intrinsic adhesion protein (e.g., integrin) properties to cluster in order to form focal adhesion contacts and mediate appropriate cellular responses. While randomly distributed adhesion ligands appear to support cell attachment, neither full cell spreading nor haptokinetic or chemokinetic motility was observed. Moreover, the size of ligand cluster controls strength of cell–substratum adhesion, cytoskeleton

**Table 1** Current techniques for micro- and nanoscale biomimetic surface engineering

| Method | Material | Resolution | References |
|---|---|---|---|
| Injection molding, casting, and embossing | Synthetic polymers | ~15 nm | Wilkinson et al. (1998) and Gadegaard et al. (2003) |
| Polymer demixing | Polymer blends (e.g., polystyrene/ polymethylmethacrylate) | 10 nm | Dalby et al. (2004a, b), Barbucci et al. (2003), Riehle et al. (2003), and Gadegaard et al. (2004) |
| Electron beam lithography | Silica, silicon, and polymers | <10 nm | Wilkinson et al. (2002) and Malaquin et al. (2004) |
| Colloidal nanolithography | Gold colloids | 50 nm | Arnold et al. (2004) and Wood et al. (2002a, b) |
| | Peptide-functionalized gold nanodots | 8 nm dot size | |
| | Nanopillars on quartz, silicon, and polymers | 20 nm | |
| Microcontact printing | Biomolecules (proteins and polysaccharides) and chemical compounds on a flat substrate | 100 nm | Wilkinson et al. (2002) and Michel et al. (2002a) |
| Self-assembling monolayers | Alkanethiols on gold, trichlorosilanes on glass, and silicon | <10 nm | Chen et al. (2000), Finnie et al. (2000), and Smith et al. (2004) |

reorganization, signaling, and cell motility (Maheshwari et al. 2000). There appears an optimal spacing of ~70 nm between individual ligands for integrin clustering and activation (Arnold et al. 2004). In addition to lateral organization, ligand presentation in the axis perpendicular to the cell membrane plays an important role. When RGD peptide is mobilized to the substrate with a polymer spacer, linker length affects efficiency of cell attachment, and that the optimal distance between RGD peptide and substrate has been shown to be ~3.5 nm (Shin et al. 2003).

## 4.3 *Brief Survey of Nanofabrication Techniques*

While the techniques for nanofabrication have been developing at a very rapid pace in the last decades, the use of nanomaterials in medicine and biology is quite innovative, and much research needs to be completed before we have a full understanding on nanomaterials and nanostructures. Development of nanomaterials and nanotechniques relies often on well-established progress in the electronic and optical engineering field (see Tables 1 and 2). Several methods have been developed

**Table 2** Nanofabrication methods for 3D cell environment engineering

| Method | Material | Structure size | References |
|---|---|---|---|
| Phase separation | Synthetic biodegrafable polymers (poly L-lactic acid) | 50 nm fiber diamater | Zhang and Ma (2000) |
| Electrospinning | Polyethylene oxide, polylactic acid, and polycaprolactone | 5–100 nm fiber diameter | Zhou et al. (2003), Sun et al. (2003), and Zussman et al. (2003) |
| Self-assembly of peptide amphiphiles | Synthetic peptide-amphiphile constructs | 5–8 nm fiber diameter | Hartgerink et al. (2002, 2001) and Silva et al. (2004) |
| Thermal assembly | Mineral/collagen composites | 100 nm-thick fibrils | Pederson et al. (2003) |
| Nanoparticle synthesis | Carbon nanotubes | Down to 3–5 nm diameter | Michalet et al. (2005), Berry and Curtis (2003), and Gillies and Frechet (2005) |
| | Superparamagnetic beads (iron oxide) | 20–500 nm diameter | |
| | Dendrimers | 10 nm | |

recently to produce synthetic scaffolds with nanofeatures for tissue engineering. For example, a phase separation technique is proposed to generate synthetic nanofibrous ECM, which mimics the fine fibrillar architecture of collagen (Zhang and Ma 2000). Three-dimensional negative replicas are produced from a porogen material. Polymeric materials, like PLA, are cast over porogen and thermally phase-separated to form nanofibrous matrices. The porogen material is then leached out with water to finally form the synthetic nanofibrous ECM with predesigned macroporous architectures. The diameter of the fibers ranges from 50 to 500 nm, which is similar to collagen matrix. These scaffolds can induce cells to assemble in a 3D fashion that resembles the natural cell organization in natural tissue.

Electrospinning of nanofibers is a relatively novel process that allows the continuous production of polymer fibers (polyethylene oxide, PLA, and polycaprolactone) ranging from less than 5 nm to over 1 μm in diameter (Zhou et al. 2003; Sun et al. 2003). The reduction of the diameter into the nanometer range gives rise to a set of favorable properties, including increase of the surface-to-volume ratio, variations in wetting behavior, and modifications of the release rate. The electrospinning technique relies on a high electric field-assisted assembly of nanofibers into well-ordered 3D structures (Zussman et al. 2003). The fibers from biodegradable polymers that can be aligned to create 3D matrix of parallel or periodic arrays may be useful for tissue engineering. A recent study from our laboratory demonstrates that human MSC adhesion and stretching are preferred along the direction of electrospun polycaprolactone (PCL) polymers of ~100 nm diameter (Fig. 5), revealing that the preferred orientation and alignment of the cells on the patterned substrate coincide with the fiber orientation. A cell viability test shows that nanofiber-directed cell orientation and alignment do not cause adverse cellular damage. This capability

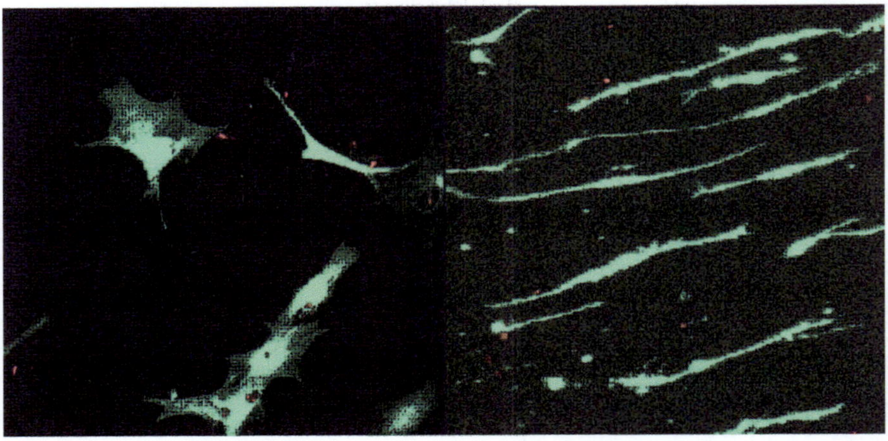

**Fig. 5** Human mesenchymal stem cells plated either on an unpatterned (*left*) or patterned (*right*) nanofibrous PCL substrate. The *white diagonal line* drawn in the *right panel* schematically illustrates the spatial orientation of the underlying PCL fibers (not fluorescently visualized). Cell viability assay indicates that cells plated on the PCL nanofibers are live (*green color*)

may be critical in engineering functional tissue by controlling the cell and ECM spatial patterns. In addition, high degree of control of cell orientation and alignment may also have significant implication for regulation of stem cell proliferation and differentiation. However, one difficulty in nanofiber technology has been seeding cells within a nanofibrillar structure with pore spaces much smaller than a cellular diameter. Somehow, the network must be formed in situ, around the cells, without cellular damage.

A class of nanofibrillar gels has been recently developed, which can self-assemble around the cells under appropriate near-physiologic conditions, and the cells survive this process (Hartgerink et al. 2002; Silva et al. 2004). The unit blocks of the 3D nanofiber matrix are peptide-amphiphile molecules, which incorporate specific biomolecular signals. The synthesized peptide-amphiphile molecule consists of long alkyl tail, amino acid spacer, and peptide sequence for a specific cell response (e.g., cell adhesion ligand RGD, laminin peptide). The nanofibers 5–8 nm diameter and up to few microns long are driven to assemble in aqueous media of high ionic strength by hydrogen bonding and hydrophobic interactions. This promising technique can potentially produce biosystems that can be delivered to living tissues by simply injecting peptide-amphiphile solution. This solution should self-assemble in vivo into artificial scaffolds directing cell differentiation, proliferation, and other crucial cell functions. The wide choice of amphiphile building-block molecules provides a versatile tool with environment-controllable and reversible self-assembly mechanism for engineering various types of tissues. For example, the self-assembled peptide-amphiphile nanofibers are shown to direct hydroxyapatite crystal nucleation and growth to form a composite material in which the alignment between hydroxyapatite and nanofibers is the same as that found in natural bone (Hartgerink et al. 2001). In-depth description for methods of nanostructure production for biological applications may be found in the later chapters.

### 4.4 Nanobioengineering

Nanomaterials could be the key component of cellular environment engineering in various applications ranging from in vitro models to practical clinical purposes. Nanofabricated materials are used in tissue engineering, neuroprostheses, orthopedic implants, and restorative medicine. For example, crystalline nanostructured (about 15 nm) hydroxyapatite coatings for dental and orthopedic implants are now commercially available (Sinha et al. 2002). These coatings with high surface area provide better osteogenic properties of the implants and significantly improve the success rate of such implants. Self-assembly on the 2D surface is a promising technique for fabrication of new generation of biocompatible synthetic biomaterials, bioselective surfaces, and even biosensors (Chaki and Vijayamohanan 2002). Self-assembling 3D scaffold systems also have a great potential to restore, maintain, or improve the tissue function. Although nanobioengineering discipline is still in its infancy, it offers potential promises for nanomaterial-based control of cell functions and responses (Zhang et al. 2002). However, extensive research effort and testing of these nanoscale systems would be required before the actual use in treatment of pathological conditions and regeneration medicine can be efficiently implemented.

Rapidly growing area of nanotechnology supplies biologists and biomedical engineers with new tools to control living cell functions and behavior via precise design of cellular environment at nanoscale level. While the fundamental knowledge is still being accumulated on how cells interact with their natural or artificial milieu, the molecular mechanisms and processes that govern the cell fate in the nanoworld remain to be elucidated. It appears clear that concentrated research effort is certainly directed to the area of nanoscale bioengineering and biosciences. Integration of nanomaterials with biotechnology to optimize extracellular environment may eventually lead to new therapies, regenerative medicine strategies, and diagnostic and bioanalytical methods.

**Acknowledgments** This work was supported in part by the grant from Office of Navy Research (N00014-06-1-0100) and a NIH grant (CA113975).

## References

Alberts B, Johnson A, Lewis J, Raff M, Roberts K, Walter P. Chapter 19: Cell junctions, cell adhesion, and the extracellular matrix. In: Molecular Biology of The Cell, 4th ed, Garland Science, Taylor & Francis Group, New York, NY 2002:1090–1124.

Alhadlaq A, Elisseeff JH, Hong L, Williams CG, Caplan AI, Sharma B, Kopher RA, Tomkoria S, Lennon DP, Lopez A, Mao JJ. Adult stem cell driven genesis of human-shaped articular condyle. Annals Biomed Eng 2004; 32:911–923.

Arnold M, Cavalcanti-Adam EA, Glass R, Blummel J, Eck W, Kantlehner M, Kessler H, Spatz JP. Activation of integrin function by nanopatterned adhesive interfaces. Chemphyschem 2004; 5:383–388.

Ayad S, Boot-Handford R, Humphries M, Kadler K, Shuttleworth A. The Extracellular Matrix Factsbook, 2nd Ed. Academic Press, San Diego, CA, 1998.

Barbucci R, Pasqui D, Wirsen A, Affrossman S, Curtis A, Tetta C. Micro and nano-structured surfaces. J Mater Sci Mater Med 2003; 14:721–725.

Berry CC, Curtis ASG. Functionalisation of magnetic nanoparticles for applications in biomedicine. J Phys D Appl Phys 2003; 36:R198–R206.

Brownlee C. Role of the extracellular matrix in cell–cell signalling: paracrine paradigms. CurrOpin Plant Biol 2002; 5:396–401.

Chaki NK, Vijayamohanan K. Self-assembled monolayers as a tunable platform for biosensor applications. Biosen Bioelectron 2002; 17:1–12.

Chen CS, Ostuni E, Whitesides GM, Ingber DE. Using self-assembled monolayers to pattern ECM proteins and cells on substrates. Methods Mol Biol 2000; 139:209–219.

Cho MR. A review of electrocoupling mechanisms mediating facilitated wound healing. IEEE Trans Plasma Sci 2002; 30:1504–1515.

Cima LG, Vacanti JP, Vacanti C, Ingber D, Mooney D, Langer R. Tissue engineering by cell transplantation using degradable polymer substrates. J Biomech Eng 1991; 113:143–151.

Colwell AS, Longaker MT, Lorenz HP. Mammalian fetal organ regeneration. Adv Biochem Eng Biotechnol 2005; 93:83–100.

Curtis A, Wilkinson C. Nantotechniques and approaches in biotechnology. Trends Biotechnol 2001; 19:97–101.

Cukierman E, Pankov R, Stevens DR, Yamada KM. Taking cell-matrix adhesions to the third dimension. Science 2001; 294:1708–1712.

Cutroneo KR. Gene therapy for tissue engineering. J Cell Biochem 2003; 88:418–425.

Curtis A. Tutorial on the biology of nanotopography. IEEE Trans Nanobiosci 2004; 3:293–295.

Dalby MJ, Giannaras D, Riehle MO, Gadegaard N, Affrossman S, Curtis ASG. Rapid fibroblast adhesion to 27 nm high polymer demixed nano-topography. Biomater 2004; 25:77–83.

Dalby MJ, Pasqui D, Affrossman S. Cell response to nano-islands produced by polymer demixing: a brief review. IEE Proc Nanobiotechnol 2004; 151:53–61.

De Arcangelis A, Labouesse EG. Integrin and ECM functions: roles in vertebrate development. Trends Genetics 2000; 16:389–395.

Desmouliere A, Chaponnier C, Gabbiani G. Tissue repair, contraction, and the myofibroblast. Wound Repair Regen 2005; 13:7–12.

Engel J. Role of oligomerization domains in thrombospondins and other extracellular matrix proteins. Int J Biochem Cell Biol 2004; 36:997–1004.

Farias E, Lu M, Li X, Schnapp LM. Integrin $\alpha_8\beta_1$–fibronectin interactions promote cell survival via PI3 kinase pathway. Bioch Biophys Res Comm 2005; 329:305–311.

Feinberg MW, Jain MK, Werner F, Sibinga NE, Wiesel P, Wang H, Topper JN, Perrella MA, Lee ME. Transforming growth factor-β1 inhibits cytokine-mediated induction of human metalloelastase in macrophages. J Biol Chem 2000; 275:25766–25773.

Friess W. Collagen-biomaterial for drug delivery. Eur J Pharm Biopharm 1998; 45:113–136.

Finnie KR, Haasch R, Nuzzo RG. Formation and patterning of self-assembled monolayers derived from long-chain organosilicon amphiphiles and their use as templates in materials microfabrication. Langmuir 2000; 16:6968–6976.

Gadegaard N, Mosler S, Larsen NB. Biomimetic Polymer Nanostructures by Injection Molding. Macromol Mater Eng 2003; 288:76–83.

Gadegaard N, Dalby MJ, Riehle MO, Curtis ASG, Affrossman S. Tubes with controllable internal nanotopography. Adv Mater 2004; 16:1857–1860.

Gillies ER, Frechet JMJ. Dendrimers and dendritic polymers in drug delivery. Drug Discov Today 2005; 10:35–43.

Geiger B, Bershadsky A, Pankov R, Yamada KM. Transmembrane crosstalk between the extracellular matrix–cytoskeleton crosstalk. Nat Rev Mol Cell Biol 2001; 2:793–805.

Hartgerink JD, Beniash E, Stupp SI. Peptide-amphiphile nanofibers: a versatile scaffold for the preparation of self-assembling materials. Proc Natl Acad Sci 2002; 99:5133–5138.

Hartgerink JD, Beniash E, Stupp SI. Self-assembly and mineralization of peptide-amphiphile nanofibers. Science 2001; 294:1684–1688.

Heino J. The collagen receptor integrins have distinct ligand recognition and signaling functions. Matrix Biol 2000; 19:319–323.

Irvine DJ, Hue KA, Mayes AM, Griffith LG. Simulations of cell-surface integrin binding to nano-scale-clustered adhesion ligands. Biophys J 2002; 82:120–132.

Jones CB, Sane DC, Herrington DM. Matrix Metalloproteinases: a review of their structure and role in acute coronary syndrome. Cardiovas Res 2003; 59:812–823.

Kleinman HK, Philp D, Hoffman MP. Role of the extracellular matrix in morphogenesis. Curr Opin Biotech 2003; 14:526–532.

Labat-Robert J. Cell–matrix interactions in aging: role of receptors and matricryptins. Ageing Res Rev 2004; 3:233–247.

Lutolf MP, Hubbell JA. Synthetic biomaterials as instructive extracellular micro-environments for morphogenesis in tissue engineering. Nat Biotechnol 2005; 23:47–55.

LeGeros RZ. Properties of osteoconducive biomaterials: calcium phosphates. Clin Orthop 2002; 395:81–98.

Maheshwari G, Brown G, Lauffenburger DA, Wells A and Griffith LG. Cell adhesion and motility depend on nanoscale RGD clustering. J Cell Sci 2000; 113:1677–1686.

Malaquin L, Vieu C, Genevieve M, Tauran Y, Carcenac F, Pourciel ML, Leberre V, Trevisiol E. Nanoelectrode-based devices for electrical biodetection in liquid solution. Microelectron Eng 2004; 73–74:887–892.

McCarthy JB, Vachhani B, Iida J. Cell adhesion to collagenous matrices. Biopolymers 1996; 40:371–381.

Michel B, Bernard A, Bietsch A, Delamarche E, Geissler M, Juncker D, Kind H, Renault JP, Rothuizen H, Schmid H, Schmidt-Winkel P, Stutz R, Wolf H. Printing meets lithography: Soft approaches to high-resolution patterning. Chimia 2002; 56:527–542.

Michel R, Reviakine I, Sutherland D, Fokas C, Csucs G, Danuser G, Spencer ND, Textor M. A novel approach to produce biologically relevant chemical patterns at the nanometer scale: Selective molecular assembly patterning combined with colloidal lithography. Langmuir 2002; 18:8580–8586.

Michalet X, Pinaud FF, Bentolila LA, Tsay JM, Doose S, Li JJ, Sundaresan G, Wu AM, Gambhir SS, Weiss S. Quantum dots for live cells, in vivo imaging, and diagnostics. Science 2005; 307:538–544.

Motlagh D, Hartman TJ, Desai TA, Russell B. Microfabricated grooves recapitulate neonatal myocyte connexin43 and N-cadherin expression and localization. J Biomed Mater Res. Part A 2003; 67:148–157.

Moore K. The Scientific Basis of Wound Healing. In: Advances in Tissue Banking, vol. 5. World Scientific Publ, 2001, 379–397.

Mott JD, Werb Z. Regulation of matrix biology by matrix metalloproteinases. Curr Opin Cell Biol 2004; 16:558–564.

Mustoe TA, Pierce GF, Morishima C, Duel TF. Growth factor induced tissue repair through direct and inductive activities in a rabbit dermal ulcer model, J Clin Invest 1991; 8:694–703.

Nissen NN, Shankar R, Gamelli RL, Singh A, DiPietro LA. Heparin and heparan sulphate protect basic fibroblast growth factor from non-enzymic glycosylation. Biochem J 1999; 338:637–42.

O'Toole EA. Extracellular matrix and keratinocyte migration. Clin Exp Dermatol 2001; 26: 525–30.

Pederson AW, Ruberti JW, Messersmith PB. Thermal assembly of a biomimetic mineral/collagen composite. Biomater 2003; 24:4881–4890.

Perris R and Perissinotto D. Role of the extracellular matrix during neural crest cell migration. Mechanisms Dev 2000; 95:3–21.

Pierschbacher MD, Hagman EG, Ruoslahti E. Location of cell attachment and proteolytic fragments of the molecule. Cell 1981; 26:259–267.

Pierschbacher MD, Ruoslahti E. The cell attachment activity of fibronectin can be duplicated by small fragments of the molecule. Nature 1984; 309:30–33.

Popat KC, Desai TA. Poly(ethylene glycol) interfaces: an approach for enhanced performance of microfluidic systems. Biosen Bioelectron 2004; 19:1037–1044.

Raman R, Sasisekharan V, Sasisekharan R. Structural insights into biological roles of protein-glycosaminoglycan interactions. Chem Biol 2005; 12:267–277.

Riehle MO, Dalby MJ, Johnstone H, MacIntosh A, Affrossman S. Cell behaviour of rat calvaria bone cells on surfaces with random nanometric features. Mat Sci Engineer C-Biomim Supramol Sys 2003; 23:337–340.

Robert L. Matrix biology: past, present and future. Pathologie Biologie 2001; 49:279–283.

Sadler JE, Shelton-Inloes BB, Sorace JM, Harlen JM, Titani K, Davies EW. Cloning and characterization of two cDNA coding for human von Willebrand factor. Proc Natl Acad Sci 1985; 82:6394–6398.

Schaffer CJ, Nanney LB. Cell biology of wound healing. Int Rev Cytol 1996; 169:151–181.

Shin H, Jo S, Mikos AG. Biomimetic materials for tissue engineering. Biomater 2003; 24:4353–4364.

Simons M, Horowitz A. Syndecan-4-mediated Cellular. Signalling 2001; 13:855–86.

Silva GA, Czeisler C, Niece KL, Beniash E, Harrington DA, Kessler JA, Stupp SI. Selective differentiation of neural progenitor cells by high-epitope density nanofibers. Science 2004; 303: 1352–1355.

Sinha M, Qadri SB, Singh CP. Preparation and Application of Nanoparticles of Ultra-pure Hydroxyapatite. 10th Foresight Conference on Molecular Nanotechnology, Bethesda, MD, October 11–13, 2002.

Smith JR. The structural basis of integrin-ligand (RGD) interaction. In: Cheresh DA and Mecham RP, eds. Integrins – Molecular and Biological Responses to the Extracellular Matrix, San Diego: Academic Press, Inc, 1994:1–32.

Smith RK, Lewis PA, Weiss PS. Patterning self-assembled monolayers. Prog Surf Sci 2004; 75:1–68.

Suzuki S, Oldberg A, Hayman EG, Pierschbacher MD, Ruoslahti E. Complete amino acid sequence of human vitronectin deduced from cDNA: Similarity of cell attachment sites in vitronectin and fibronectin. EMBO J 1985; 4:2519–2524.

Sun S, Wise J, Cho MR. Human fibroblast migration in 3D collagen gel in response to non-invasive electrical stimulus. I: Characterization of induced 3D cell movement. Tissue Eng 2004; 10:1548–1557.

Sun S, Cho MR. Human fibroblast migration in 3D collagen gel in response to non-invasive electrical stimulus. II: Identification of electrocoupling molecular mechanisms. Tissue Eng 2004; 10:1158–1565.

Sun ZC, Zussman E, Yarin AL, Wendorff JH, Greiner A. Compound core-shell polymer nanofibers by co-electrospinning. Adv Mater 2003; 15:1929–1932.

Svetlana A. Kuznetsova, Robert DD. Functional regulation of T lymphocytes by modulatory extracellular matrix proteins. Int J Biochem Cell Biol 2004; 36:1126–1134.

Takoaka K, Nakahara H, Yoshikawa H, Masuhara T, Tsuda T, Ono K. Ectopic bone induction on and in porous hydroyapatite combined with collagen and bone morphogenetic protein. Clin Orthop 1998; 234:250–254.

Tao SL, Lubeley MW, Desai TA. Bioadhesive poly(methyl methacrylate) microadhesives for controlled drug delivery. J Controlled Release 2003; 88:215–228.

Tan W, Desai TA. Layer-by-layer microfluidics for biomimetic three-dimensional structures. Biomater 2004; 25:1355–1364.

Trelstad R. The extracellular matrix in development and regeneration. Int J Dev Biol 2004; 48:687–694.

Tuan TL, Song A, Chang S, Younai S, Nimni ME. In vitro fibroplasia: matrix contraction, cell growth, and collagen production of fibroblasts cultured in fibrin gels. Exp Cell Res 1996; 25:127–134.

Wahl SM. Transforming growth factor beta. In: Gallin J and Snyderman R, eds. Inflammation: Basic Principles and Clinical Correlates. Lippincott-Raven Publishers, Philadelphia, 1999, 883–892.

Wilkinson CDW, Riehle M, Wood M, Gallagher J, Curtis ASG. The use of materials patterned on a nano- and micro-metric scale in cellular engineering. Mat Sci Engineer C 2002; 19:263–269.
Wilkinson CDW, Curtis ASG, Crossan J. Nanofabrication in cellular engineering. J Vac Sci Technol B 1998; 16:3132–3136.
Wood MA, Meredith DO, Owen GR. Steps toward a model nanotopography. IEEE Trans Nanobiosci 2002; 1:133–140.
Wood MA, Riehle M, Wilkinson CDW. Patterning colloidal nanotopographies. Nanotech 2002; 13:605–609.
Yannas IV. Tissue regeneration templates based on collagen-glycosaminoglycan copolymers. Adv Polymer Sci 1995; 122:219–244.
Yang SH, Leong KF, Du ZH, Chua CK. The design of scaffolds for use in tissue engineering. Part I. Traditional factors. Tissue Eng 2001; 7:679–689.
Yoneda A, Couchman JR. Regulation of cytoskeletal organization by syndecan transmembrane proteoglycans. Matrix Biol 2003; 22:25–33.
Yoon YW, Kwon HM, Hwang KC, Choi EY, Hong BK, Dongsoo Kim D, Kim HS, Cho SH, Song KS and Sangiorgi G. Upstream regulation of matrix metalloproteinase by EMMPRIN; extracellular matrix metalloproteinase inducer in advanced atherosclerotic plaque. Atheroscler 2005; 180:37–44.
Yurchenco PD, Birk DE. Extracellular Matrix Assembly and Structure, Academic Press, San Diego, CA, 1994.
Zagris N. Extracellular matrix in development of the early embryo. Micron 2001; 32:427–438.
Zhang S, Marini DM, Hwang W, Santoso S. Design of nanostructured biological materials through self-assembly of peptides and proteins. Curr Opin Chem Biol 2002; 6:865–871.
Zhang R, Ma PX. Synthetic nano-fibrillar extracellular matrices with predesigned macroporous architectures. J Biomed Mater Res 2000; 52:430–438.
Zhou YX, Freitag M, Hone J, Staii C, Johnson AT, Pinto NJ, MacDiarmid AG. Fabrication and electrical characterization of polyaniline-based nanofibers with diameter below 30 nm. Appl Phys Lett 2003; 83:3800–3802.
Zipfel WR, Williams RM, Christie R, Nikitin AY, Hyman BT, Watt WW. Live tissue intrinsic emission microscopy using multiphoton-excited native fluorescence and second harmonic generation. Proc Natl Acad Sci 2003; 100:7075–7080.
Zoumi A, Yeh A, Tromberg BJ. Imaging cells and extracellular matrix in vivo by using second-harmonic generation and two-photon excited fluorescence. Proc Natl Acad Sci 2002; 99:11014–11019.
Zussman E, Theron A, Yarin AL. Formation of nanofiber crossbars in electrospinning. Appl Phys Lett 2003; 82:973–975.

# Part II
# Synthesis and Characterization Approaches

# Synthesis and Patterning Methods for Nanostructures Useful for Biological Applications

**Chiara Daraio and Sungho Jin**

## 1 Introduction

The importance and benefits of nanotechnology in biology and medicine are now well-recognized by scientists, technologists, as well as various governmental and private research funding agencies. The basis that enables the application of nanotechnology is the availability of nanostructured materials. Therefore, it is essential to provide an insight on the state-of-the-art methods to manufacture various nanostructures.

Nanotechnology has exploded onto the scientific scene in the last 20 years or so after the discovery, followed by a Nobel Prize recognition, of novel forms of carbon as the fullerenes molecules or buckyballs (Kroto et al. 1985), and the rediscovery of carbon nanotubes (CNTs) (Iijima 1991). The subsequent development has impacted nearly every field of scientific research. The infancy of nanotechnology began in materials science laboratories and has now matured at the forefront of several other disciplines, including chemistry, physics, various engineering fields, and, of course, biology and biomedical engineering. Nanotechnology is now rapidly being expanded for improved biotechnology to the extent that it can now be called as a subfield of "nanobiotechnology." The research and development on synthesis of nanowires and nanoparticles, nanopatterning, as well as nanoscale phenomena and properties literally exploded in the past several years developing a large variety of nanostructures extending to a myriad of other physical or chemical components. Some of the biological applications of the nanostructures, such as nanopatterns, nanochannels, nanotubes, and nanoparticles, synthesized using the

C. Daraio
Division of Engineering and Applied Science, California Institute of Technology, Pasadena, CA, USA

S. Jin (✉)
Materials Science & Engineering Program, Mechanical and Aerospace Engineering Department, University of California, San Diego, La Jolla, CA, USA

G.A. Silva and V. Parpura (eds.), *Nanotechnology for Biology and Medicine: At the Building Block Level*, Fundamental Biomedical Technologies, DOI 10.1007/978-0-387-31296-5_2, methods described in this chapter include nanoparticles for fluorescent biological labeling and tagging, drug and gene delivery, biodetection of pathogens and proteins, nanotubes and nanochannels for probing the structure of DNA, tissue engineering, magnetic nanoparticles for tumor destruction via heating (hyperthermia) and magnetic resonance imaging contrast enhancement, separation and purification of biological molecules and cells, and many others that are still currently under investigation (Salata 2004).

The methods for synthesis and patterning for many of these nanostructures are founded on basic, well-known techniques and their modifications, as is described in this chapter. The nanofabrication processes can be divided into the two well-known approaches: "top-down" and "bottom-up." The "top-down" approach uses traditional methods to guide the synthesis of nanoscale materials. The paradigm proper of its definition generally dictates that in the "top-down" approach it all begins from a bulk piece of material, which is then gradually or step-by-step removed to form objects in the nanometer-size regime (~1 billionth of a meter or $10^{-9}$ m). Well-known techniques, such as photo lithography and electron beam lithography, anodization, and ion- and plasma etching (PE), that are later described, all belong to this type of approach. The top-down approach for nanofabrication is the one first suggested by Feynman in his famous American Physical Society lecture in 1959, "There is plenty of room at the bottom" (Feynman 1959):

> Now comes the interesting question: How do we make such a tiny mechanism? I leave that to you. However, let me suggest one weird possibility. You know, in the atomic energy plants they have materials and machines that they can't handle directly because they have become radioactive. To unscrew nuts and put on bolts and so on, they have a set of master and slave hands, so that by operating a set of levers here, you control the "hands" there, and can turn them this way and that so you can handle things quite nicely.
>
> Most of these devices are actually made rather simply, in that there is a particular cable, like a marionette string, that goes directly from the controls to the "hands." But, of course, things also have been made using servo motors, so that the connection between the one thing and the other is electrical rather than mechanical. When you turn the levers, they turn a servo motor, and it changes the electrical currents in the wires, which repositions a motor at the other end.
>
> Now, I want to build much the same device – a master-slave system which operates electrically. But I want the slaves to be made especially carefully by modern large-scale machinists so that they are one-fourth the scale of the "hands" that you ordinarily maneuver. So you have a scheme by which you can do things at one-quarter scale anyway – the little servo motors with little hands play with little nuts and bolts; they drill little holes; they are four times smaller. Aha! So I manufacture a quarter-size lathe; I manufacture quarter-size tools; and I make, at the one-quarter scale, still another set of hands again relatively one-quarter size! This is one-sixteenth size, from my point of view. And after I finish doing this I wire directly from my large-scale system, through transformers perhaps, to the one-sixteenth-size servo motors. Thus I can now manipulate the one-sixteenth size hands. Well, you get the principle from there on. It is rather a difficult program, but it is a possibility …

The "bottom-up" approach on the other hand takes the idea of "top-down" approach and flips it right over. In this case, instead of starting with large materials and chipping them away to reveal small bits of them, it all begins from atoms and molecules that get rearranged and assembled to larger nanostructures. It is the new paradigm for synthesis in the nanotechnology world as the "bottom-up" approach allows a creation of diverse types of nanomaterials, and it is likely to revolutionize the way we make materials. It requires a thorough understanding of the short-range forces of attraction, such as Van der Waals forces, electrostatic forces, and a variety of interatomic or intermolecular forces. Since it is not possible to have various minute things come together without some attractive force or active field of force in the region, having the fundamental forces "doing all the work" for us is the key principle underlying this approach.

Some examples of such a synthesis route starting from atoms and molecules are methods like self-assembly of nanoparticles or monomer/polymer molecules, chemical or electrochemical reactions for precipitation of nanostructures, sol–gel processing, laser pyrolysis, chemical vapor deposition (CVD), plasma or flame spraying synthesis, and atomic or molecular condensation. The bio-assisted synthesis of nanomaterials also belongs to this approach (Sarikaya et al. 2003). However, despite being so promising and inviting, our ability to build things from the bottom up is fairly limited in scope. While we can assemble relatively simple structures, we cannot produce complex, integrated devices using the bottom-up approach. Any kind of overall ordered arrangement aside from repeating regular patterns cannot be done without some sort of top-down influence, like lithographic patterning. Until we have fully mastered the bottom-up synthesis approach, we will not be able to fully exploit its speed and accuracy. The important factor is that they are two different approaches to creating nanostructures which can be applied according to the specific needs for each application, often in a complementary way.

## *1.1 Nanofabrication by Top-Down Methods*

Top-down methods were originally introduced as a new set of manufacturing tools to solve specific industrial problems in micromachining, such as the microelectromechanical systems (MEMS) devices. At the nanoscale though, these methods are generally not suitable for production on a very large scale because they presently encounter technological limitation and require extremely long and costly processes. Nonetheless, they represent some of the most common approaches now used to pattern surfaces and create three-dimensional (3-D) features on substrates. Over the past decades, there has been a wide variety of top-down manufacturing techniques that have been implemented using different media that span from chemical and electrochemical means to photofabrication, laser machining, electron- and ion-beam milling (IBM), plasma etching and powder blasting. Here, we describe only the most common techniques used for nano- or nanobio-related applications.

### 1.1.1 Nanolithography

Nanolithography is one of the most established techniques for making nanostructured materials and patterns. Nanolithography derives its name from the Greek words *nanos* (dwarf), *lithos* (rock), and *grapho* (to write), which literally means "small writing on rocks." This technique is based on depositing, etching, or writing

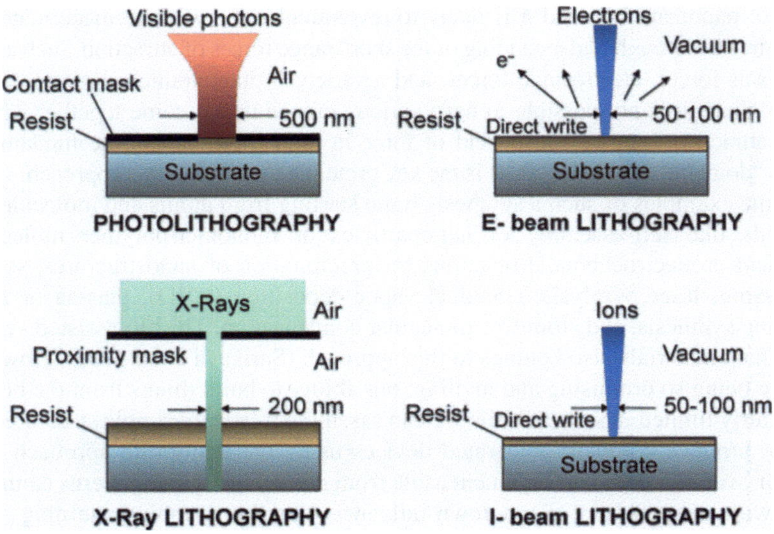


**Fig. 1** Comparison of the available lithography techniques (image adapted from Brodie and Muray 1982)

on a surface features with dimensions on the order of nanometers. Lithography (Madou 2002) can be performed using light (optical- or photolithography), electrons (e-beam lithography), ions (i-beam lithography), or X-ray (X-ray lithography, LIGA) depending on the desired minimum feature size of the outputs (Fig. 1).

Photolithography is the most common and widely used technique in applications requiring micron and submicron feature sizes, like electronic, integrated circuits manufacturing, and microfluidic devices. It finds a lot of applications in bio-MEMS devices for its applicability to patterning of organic and inorganic materials, hydrogels, membranes, and ion-selective electrodes (Lambrechts 1992). With this technique, a design or pattern can be transferred at once on the surface of a device by exposure with light. Similarly, e-beam, i-beam, and X-ray lithography use the exposure to beams of electrons, ions, or X-ray for the design transfer.

All the lithographic techniques are multistep processes. As an example, we describe the steps followed in the specific case of photolithography of a Si/$SiO_2$ substrate (Fig. 2), keeping in mind that other techniques follow a very similar path. The first step is to coat the surface of the desired substrate with a thin polymer layer of a positive or negative photoresist (Fig. 2a), which is then covered by a predesigned, light-blocking mask and exposed to UV light (Fig. 2b). After exposure, the cured photoresist remains on the substrate as a protective coating, and the uncured portions are removed by the developer (Fig. 2c). This step is then followed by the etching of the substrate and the final removing of the photoresist (Fig. 2d). The type of resist chosen allows having a direct duplication of the mask used (as in the case of positive resist) or inverse of it (negative resist) exactly like in the case of film photography developing.

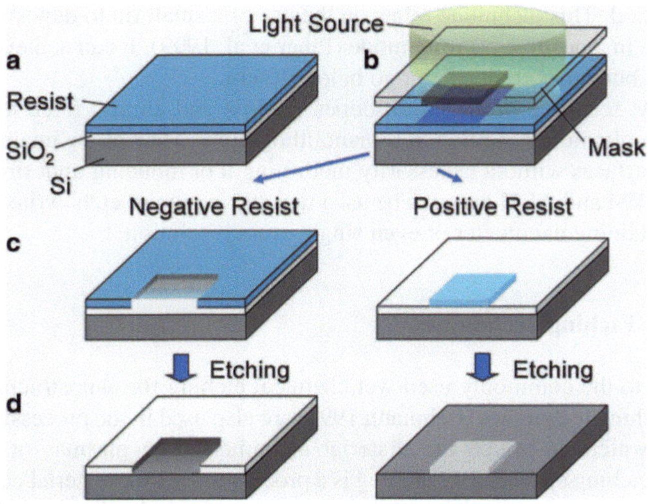


**Fig. 2** Photolithography steps of surface patterning

The technical limits of photolithography are related to the wavelength of light used, which also determines the smallest feature sizes achievable. A limitation is dictated by the lens characteristics: if the used wavelength is made too small, the lens simply absorbs the light in its entirety and do not allow curing of the photoresist. This implies that photolithography cannot reach the superfine sizes of some other technologies, such as electron beam lithography, as the wavelength of typical light, even UV lasers, is somewhat limited. Small-wavelength UV light (deep UV techniques) can now be employed to obtain finer nanostructures.

A technology that allows for fabrication of much smaller nanostructures than photolithography is that of electron beam (e-beam) lithography. The use of an electron beam to draw a pattern nanometer by nanometer and expose an e-beam resist layer (usually made of a polymer) allows achieving incredibly small sizes (on the order of 20 nm). This technique, though, is much more expensive and time consuming than photolithography, as the e-beam exposure is carried out line-by-line or dot-by-dot, unlike the photolithography by which the whole area of, e.g., 12-in. diameter Si wafer can be exposed all at once. It takes orders of magnitude longer to complete the patterning with e-beam, so it is therefore difficult to sell the e-beam lithography for large-scale industry applications.

New nanolithography technologies are continued to being researched, leading to smaller possible sizes and access to more complex surface geometries and properties. Extreme ultraviolet (EUV) lithography, for example, is capable of using light at wavelengths of 13.5 nm and fabricating ~30-nm regime features. While hurdles still exist in this new field, it promises the possibility of sizes far below those produced by current industry standards. Another lithography technique proposed for the realization of nanoscale features is the direct-write, "dip-pen" nanolithography

(DPN) method. This technique relies on the use of a small tip to deposit molecules on a surface in a positive printing mode (Piner et al. 1999). It can achieve relatively small sizes, but currently it cannot go below 40 nm.

Similarly, scanning probe microscopes (SPMs) and atomic force microscopes (AFMs) are often used in relation to nanolithography. They allow imaging the fine details on surfaces without necessarily modifying it or inducing undesired damage. However, SPM and AFM can also be used to achieve precise etch, write, or print on surfaces obtaining nanometer or even single-atom resolution.

### 1.1.2 Dry Etching Techniques

In addition to the commonly used wet chemical etching for nanostructure fabrication, dry etching techniques (Lehmann 1991) are also used in the processing of semiconductor wafers in which the material is immersed in plasmas or discharges containing etching species. Dry etching is a process to remove material carried out in the gas phase that uses no aqueous solutions, such as an acid or base, hence the name "dry etching." Dry etching techniques can be either purely chemical (plasma etching), purely physical (ion beam milling, IBM), or a combination of both (reactive ion etching, RIE). The dry etch reactions are generally isotropic, taking place with equal rate in any direction. They can be very selective etching techniques for removal of one material relative to another, generally used in those applications in which directionality (anisotropy) of etching in not required, e.g., in resist stripping.

Plasma etching, like most dry etching techniques, is carried out in chambers in which high-energy electric and magnetic fields rapidly dissociate suitable feeding gasses, forming a plasma. When the surface of a material gets in contact with the plasma, it gets bombarded by molecules, ions, electrons, and photons which stimulate the expulsion of material from the surface, leading to atomically precise removal of materials. Commonly, plasmas can be created in different ways: using direct current (DC) electric fields, applying a radio frequency (RF) voltage source or using a microwave source. One of the most common plasma etching techniques used is RIE because of its relative simplicity and versatility. The RIE method requires process pressures varying between $10^{-3}$ and $10^{-1}$ Torr. A schematic diagram summarizing the differences between some of the dry etching methods is shown in Fig. 3 (Madou 2002).

Another common dry etching technique is IBM. In this case, the plasma source is decoupled from the substrate, which is placed on a separate third electrode in a triode setup. IBM can be employed to cover relatively large areas. A commonly used IBM method is focused ion beam (FIB), usually coupled to a high-resolution scanning electron microscope (SEM) to provide visual guidance during the writing process. FIB uses a beam of ions (usually $Ga^{2+}$) that gets narrowed in a very small area and it is employed as a direct writing tool for patterning holes, thinning samples, and depositing insulating or conductive material almost atom by atom. FIB milling is a powerful technique, now commonly considered a "mechanical" machining technique in which the classical metal drill bit used in all machine shops is

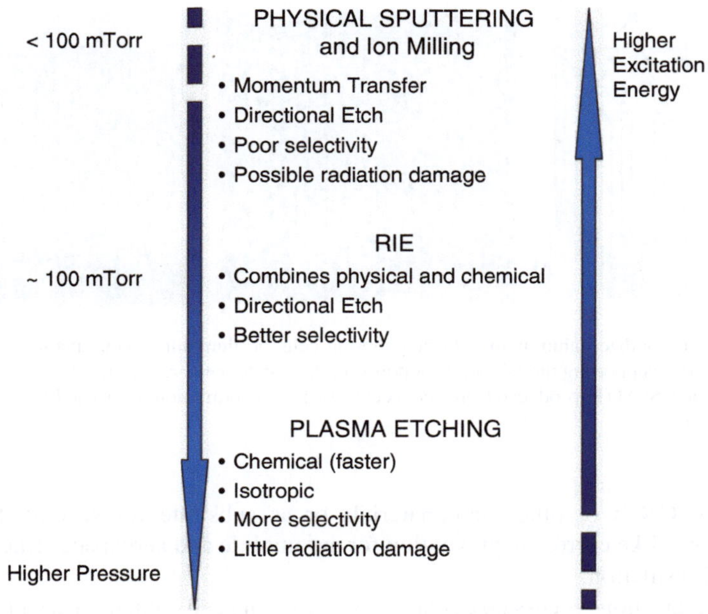


**Fig. 3** Schematic diagram of the dry etching techniques spectrum (modified from Madou 2002)

replaced by the focused beam of highly energetic ions. However, the disadvantages of FIB are long processing time, the risk of undesired ion implantation, or damage to the sample.

### 1.1.3 Anodization

Anodization is a term that stands for anodic oxidation and it is "the process of growth of an oxide on the surface of a biased conductive solid (e.g., semiconductor wafer) immersed in a liquid electrolyte or gas. With this method, under specific process conditions, a thin, dense, barrier oxide of uniform thickness can be grown on several different metals" (Alwitt 2002). The properties of the grown oxide layer vary as a function of the materials selected and the process conditions. The commercial application of anodic oxidation extends to different surface coatings, mostly based on aluminum, tantalum, niobium, and zirconium. Recently, titanium has been also receiving increasing attention in bio-related applications (Oh et al. 2005). Aluminum in particular is commonly used because of its ease and versatility of processing. In addition to the creation of a thin barrier oxide, the anodization of aluminum, under specific conditions, can produce a thick oxide coating with a high density of vertically oriented nanopores (see Fig. 4) (Asoh et al. 2001). This porous substrate can then function as template for other assisted

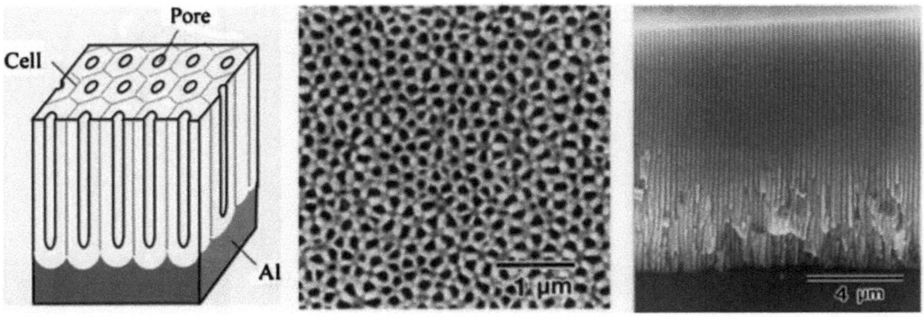


**Fig. 4** Porous-anodized aluminum with neighboring cells of aluminum oxide merges together to form the matrix supporting the aligned nanopores. *Left*, scheme; *center*, top view transmission EM; *right*, side view SEM (Reproduced from Asoh et al. 2001 with permission from The Electrochemical Society, Inc.)

growth of CNTs or other nanomaterials or as substrate for several different applications, like corrosion prevention for automobile and aerospace structures or electrical insulation.

The anodization process takes place in a processing cell containing a cathode and an anode. Usually, the metallic sample of choice functions as the anode, and it is connected to the positive terminal of a DC power supply. A different conductive material, inert in the anodizing bath, functions as the cathode (some example materials used as cathodes are carbon, lead, nickel, or stainless steel), and it is connected to the negative terminal of the supply. When this circuit is closed, electrons are allowed to move, being attracted by the positive terminal, and permit the reaction between the ions present on the surface of the metal with the water molecules that forms the oxide layer.

## *1.2 Nanofabrication by Bottom-Up Methods*

With the first decades of the new millennium, standard top-down techniques for nanomanufacturing have approached a critical limit in the ability to miniaturize components. Further size reduction has required innovative chemical routes. Eric Drexler, a pioneer in nanotechnology, in 1986 said, "our ability to arrange atoms lies at the foundation of technology." From this point on, several bottom-up approaches have blossomed, trying to build nanostructures one atom at a time, in the most efficient and precise way possible. On this aspect, nature is well-ahead of human engineering: proteins build structures quickly and effectively mastering the self-assembly of molecules as the basic building blocks of life. Nanoengineering is today putting a lot of effort in combining old, well-known chemical methods with new systems, even looking for inspiration from nature in bio-inspired structures and bio-assisted synthesis of nanodevices (Madou 2002).

One of the basic bottom-up techniques used to fabricate nanomaterials is chemical precipitation. This method allows producing nanoparticles of metals, alloys, oxides, etc. from aqueous or organic solutions following relatively simple and cost-effective steps. The drawback of this synthesis method, however, is the difficulty in the ability to control the distribution of particle size and shape, which limits the use of the produced nanomaterials in applications where a random particle distribution is undesired. Nanoscale precipitates can be obtained with various techniques, for example by a controlled phase transformation guided by the free energy diagrams or by controlling the solid-state diffusion, following a composite route approach, for example mixing two different materials and stirring them mechanically. Other approaches can be found in exploiting internal oxidation of materials, in thin film deposition of coatings, or sputtering.

The phase transformation approaches rely mostly on the rapid cooling of a solution that, not given enough time to reestablish equilibrium transformations, becomes supersaturated and precipitates into more stable and fine solid states. Common methods for rapidly cooling a solution (achieving cooling rates between $10^5$ and $10^8$ degrees per second) are splat cooling, quenching in salt baths, or melt spinning. These techniques are often used to obtain thin films and bulk materials with an extremely fine microstructure that influences positively its mechanical and electrical properties. Another common example of nanofabrication of nanoparticles or nanostructured films based on phase transformation of materials is called "spinodal decomposition." It is a spontaneous reaction driven by the free energy minimization of the material that guarantees the formation of a periodic structure of nanoparticles of equal sizes.

In order to control or limit the growth of nanomaterials produced with bottom-up methods, two techniques are conventionally used: arrested precipitation and physical restriction. Arrested precipitation relies on the exhaustion of one of the reactants or on the introduction of chemicals that would block the reaction. The physical restriction method relies on the confinement of the volume available during growth of the individual nanoparticles. Example of physical restriction includes using the micelle-reverse-micelle reactions or using templates (like, for example, the porous alumina described in the Sect. 1.1.3 above).

There are numerous bottom-up methods used to produce nanomaterials; we limit the rest of the chapter to the description of a few: sol–gel processing, CVD, self-assembly and bio-assisted synthesis, laser pyrolysis, electroplating, plasma or flame spraying synthesis, atomic or molecular condensation, and supercritical fluid synthesis.

### 1.2.1 Sol–Gel Processing

The sol–gel process in general is based on the transition of a system from a liquid "sol" (mostly a colloidal suspension of particles) into a gelatinous network "gel" phase. With this, it is possible to create at low temperature ceramic or glass materials in a wide variety of forms. It is a long-established industrial process that is

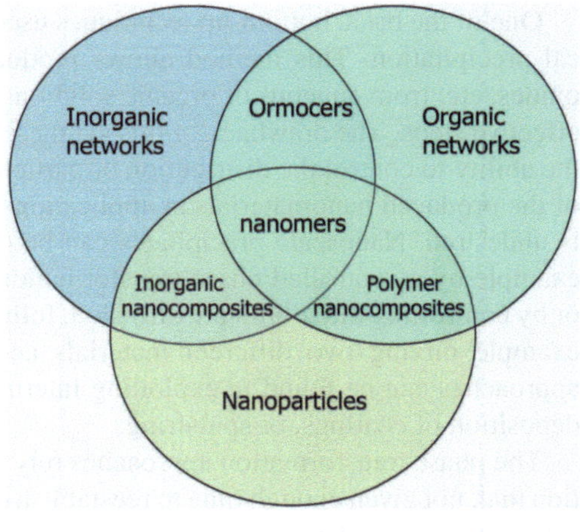


**Fig. 5** System model for nanocomposites produced by sol–gel (modified from Luther 2004)

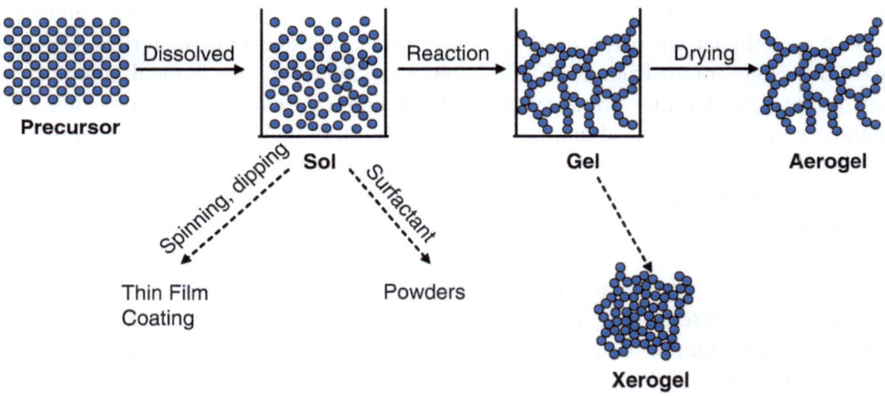


**Fig. 6** Basic flow of a sol–gel process (modified from Madou 2002)

very cost-effective and versatile. It has been further developed in last year for the production of advanced nanomaterials and coatings but also in bio-MEMS application for the production of piezoelectrics, such as lead–zirconium–titanate (PZT) (Madou 2002) or membranes (Guizard et al. 1992). Sol–gel processes are well-adapted for oxide nanoparticles (Lakshmi et al. 1997) and composite nanopowder synthesis as well as for access to organic–inorganic materials. A summary of all the possible interlinked combination of organic and inorganic nanocomposites that are produced by the sol–gel method is provided in Fig. 5.

The basic flow of the sol–gel process is described in Fig. 6. The "sol" is prepared by mechanically mixing a liquid alkoxide precursor (such as tetramethoxysilane, TMOS, or tetraethoxysilane, TEOS), water, a cosolvent, and an acid or base catalyst at room temperature. During this step, the alkoxide groups are removed by the acid- or base-catalyzed hydrolysis reactions, and networks of O–Si–O linkages are formed in subsequent condensation reactions. After this step, the treatment of the sol is varied depending on the final products desired. For example, spinning or dipping techniques can create thin film coating, and the exposure of the sol to a surfactant can lead to powders. Depending on the water–alkoxide molar ratio $R$, the pH, the temperature, and the type of solvent chosen, additional condensation steps can lead to different polymeric structures, such as linear, entangled chains, clusters, and colloidal particles. In some cases, the resulting sol is cast into a mold and dried to remove the solvent. This leads to the formation of a solid structure in the shape of the mold (e.g., aerogels and xerogels) with large surface-to-volume ratios, high pore connectivity, and narrow pore size distribution. They can be doped with a variety of organic/inorganic materials during the mixing stage to target specific applications.

### 1.2.2 Chemical Vapor Deposition

The CVD process is now probably the most common of all bottom-up approaches. It is used today to grow structures, like nanotubes, nanowires, and nanoparticles aided by several different types of chambers and growth-enhancing methods. The process consists of decomposing a gaseous precursor that adheres and accumulates onto a substrate (i.e., a silicon wafer or a quartz slide) (Jensen 1989). The presence of a catalyst, either predeposited on the substrate or provided in the gas feedstock, activates the chemical reaction between the substrate surface and the gaseous precursor. The CVD reaction can be achieved either with temperature (thermal CVD) or with plasma (PECVD). Plasmas can be obtained with DC electric fields, RF fields, or microwave fields, and their presence allows decreasing significantly the process temperature compared to the thermal CVD process. The presence of plasma also enables a more aligned or directional growth of the desired nanomaterials. A number of forms of CVD systems are in wide use and are frequently referenced in the literature. Some examples include atomic-layer CVD (ALCVD) in which two complementary precursors [e.g., $Al(CH_3)_3$ and $H_2O$] are alternatively introduced into the reaction chamber; metal-organic CVD (MOCVD) in which metal-organic precursors are used to obtain specific crystalline structures [e.g., tantalum ethoxide, $Ta(OC_2H_5)_5$, to create TaO nanostructures, and tetra dimethyl amino titanium (TDMAT) to create TiN]; laser-assisted CVD (LCVD); rapid thermal CVD (RTCVD) that uses heating lamps or other methods to rapidly heat the wafer substrate; ultrahigh vacuum CVD (UHVCVD); and more (Adams 1988; Jensen 1989).

The CVD process is widely used to produce CNTs and semiconductor nanowires, such as Si, GaN, and ZnO. CNTs have received much attention in the recent years for their potential application in several fields of bioengineering, from enhanced cell growth to biosensing, biomanipulation, and drug delivery. The most common synthesis routes for CNTs are CVD processes because they allow large-scale production of CNTs with high purity and good yield. To clarify how the CVD process works, we show a schematic diagram of a thermal CVD growth system used for the synthesis of CNTs (Fig. 7). In this example, a carbon-containing precursor is

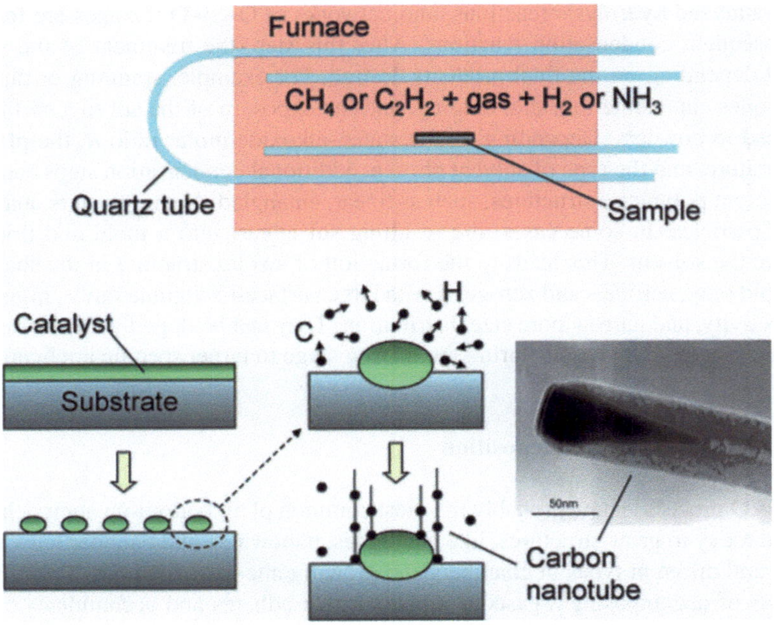


**Fig. 7** Schematics of a thermal CVD furnace (*top*) and basic flow of a carbon nanotube growth process (*bottom*)

decomposed in the furnace at high temperature. The growth reaction is activated by the presence of a catalyst that can be deposited on the sample's surface (as in the figure) or in the gas mixture fed to the reactor (e.g., injecting ferrocene $C_{10}H_{10}Fe$). The most common carbon sources used in thermal CVD growth of CNTs are mixtures of ammonia ($NH_3$) and acetylene ($C_2H_2$) while the catalyst metal can be cobalt, iron, or nickel. The growth proceeds through several steps: first, the catalyst metal breaks up into islands at high temperatures and forms metal seeds for the reaction; then, the hydrocarbon gas supplied in the quartz tube decomposes creating floating C and H atoms; the floating atoms are then attracted by the catalyst seeds that become supersaturated and condensate forming ordered tube-shaped graphene sheets. The catalyst particle can then either stay attached to the substrate (base growth) or get pushed to the tip of the formed nanotube (tip growth) conferring different properties to the final structure. This type of growth process, although being relatively inexpensive, forms a randomly distributed array of CNTs with a complex hierarchical microstructure. Bundles of tubes grow vertically, but the tubes are intertwined and partially curled at small scales. This lack of complete alignment is normally overcome by the use of PECVD systems, which are a bit more complex and expensive, since they need additional constituents like electrodes, pumps, voltage supplies, etc.

### 1.2.3 Self-Assembly and Bio-Assisted Synthesis

As mentioned in the earlier section for the case of CNT growth, the assembly of nanostructures or nanoparticles in a periodically aligned fashion or in a functionally engineered geometry is often desired over random or disordered structures. Approaches for ordering or assembling nanostructures in a desired geometry have been widely investigated in the past and are still pursued in many different laboratories. Self-assembly provides the most convenient way for alignment, especially considering that at such small scales, single-structure manipulation and postgrowth rearrangement appear to be very expensive if not at all impractical. Chemistry and molecular biomimetics are the dominant fields leading the design of hybrid technologies combining the tools of molecular biology and nanotechnology. Molecular biomimetics combines the physical and biological fields to assemble nanomaterials using the recognition properties of proteins.

Researchers have found ways to process nanoparticles of a wide range of materials – including organic and biological compounds, inorganic oxides, metals, and semiconductors – using chemical self-assembly techniques. In these cases, molecules are attached, patterned, or clustered on specific substrates or to themselves using chemical and biomolecular recognition (e.g., the preferential docking of DNA strands with complementary base pairs). Other techniques, like micelle, reverse micelle, and photochemical and sonochemical synthesis, are also employed to realize one-, two-, and three-dimensional self-assembled nanostructures. Lately, viruses have also been used to assemble specific nanostructures (Mao et al. 2004). A review of the path to nanotechnology through biology is provided in (Sarikaya et al. 2003). It was shown that proteins, through their specific interactions with inorganic and other macromolecules, could be used in nanotechnology to control structures and functions just like they do biological tissues in organisms. Taking lessons from nature, polypeptides can be genetically engineered to specifically bind to selected inorganic compounds for specific applications, as outlined in Fig. 8 (Sarikaya et al. 2003).

The use of organic molecules is also a powerful technique in the synthesis and arrangement of semiconducting nanoparticles, for example quantum dots, used in biological labeling and tagging. One of the major issues in the synthesis of these isolated islands of materials is to prevent particles' agglomeration and coarsening. Micelle processes and organic ligands capping (TOPO) are synthesis methods based on the simple mixing of surfactant with desired materials' sources. With these methods, it is possible to synthesize and/or align a wide range of nanomaterials spanning from magnetic nanoparticles (i.e., $Fe_3O_4$ particles found in magnetotactic bacteria) to the bright CdS quantum dots used for bioimaging. Micelles are employed as small chemical-reacting chambers in which nanomaterials are synthesized in fixed-sized cells (Fig. 9). One of the first examples of nanoparticle synthesis using a reverse-micelle process was for fabrication of FePt nanoparticles (Sun et al. 2000) using a combination of oleic acids and oleyl amine to stabilize the monodisperse FePt colloids and to prevent oxidation. With this method, the authors showed that it is also possible to control the size of the synthesized nanoparticles and their composition,

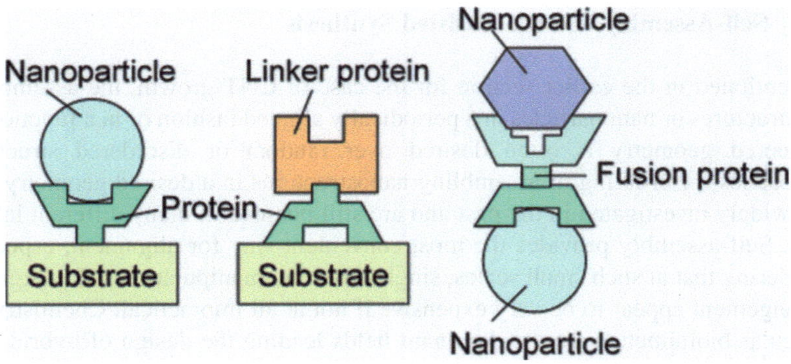


**Fig. 8** Examples of molecular biomimetics (adapted from Sarikaya et al. 2003). Proteins can be used, for example, to bind nanoparticles to a functionalized substrate, to create linkers onto a specific substrate, or to connect different nanoparticles with each other

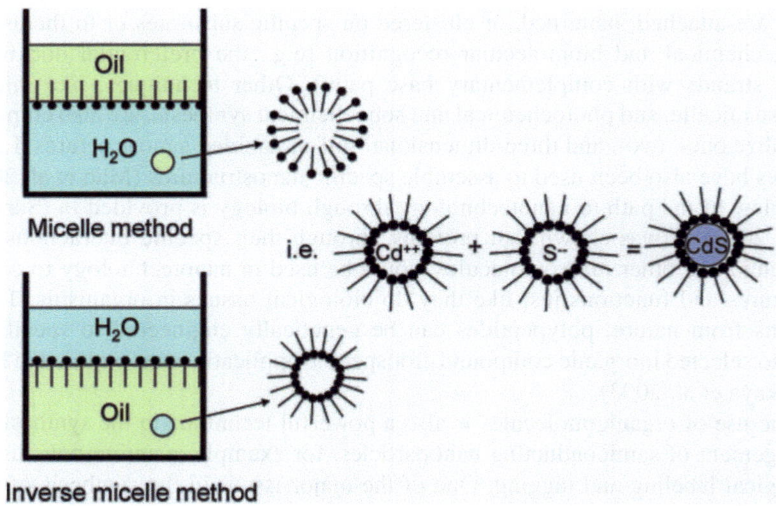


**Fig. 9** Micelle and reverse micelle processing for synthesis of nanoparticles. On the *right*, it is shown the schematic for the synthesis of CdS quantum dots via reverse (inverse) micelle method

simply tuning the molar ratio of the solvents. Upon drying, the particles could self-assemble into three-dimensional superlattices that are chemically and mechanically robust and can, for example, support high-density magnetization reversal transitions.

The micelle process is also commonly used in the synthesis of semiconducting quantum dots. For example, CdSe nanoparticles capped with ZnS coating offer significant advantages as bioimaging agents as compared to the commonly used organic fluorescent dye materials. Quantum dots are ~20 times brighter than the typical organic dye molecules, and their emission is ~100 times more stable to photobleaching than conventional fluorescent dyes (e.g., rhodamine). They can be made water soluble by chemical functionalization and can also be made specific cell- or organ-targetable via chemical or biological conjugation (Chan and Nie 1998; Akerman 2002). They can be imaged with a conventional fluorescence microscope. Besides the visible-emitting quantum dots, both near-IR and UV-emitting dots can also be synthesized.

The essence of bottom-up processing of nanotechnologies is mastered by nature in biological systems. For example, cells function on similar length scales to those of synthetic nanosystems. The molecules that operate and react in living cells are similar to the devices imagined for building nanosystems. Rather than using biological elements for the assembling of inorganic nanostructures, another way proposed for the synthesis of nanomaterials relies on incorporating the biological components themselves into artificial nanoscale structures and devices. Biological molecular motors, for example, have been incorporated into artificial structures, and the light harvesting complexes of plants or photosynthetic bacteria have been incorporated into synthetic membranes. From this point on, a vast array of applications that combine artificial structures with biological components can be envisioned and investigated, defining the future of nanobiotechnology.

### 1.2.4 Laser Pyrolysis

Laser pyrolysis is a technique that uses IR laser energy to decompose materials by heating gas-phase reactants in an oxygen-deprived environment. This method is used to produce extremely pure spherical nanoparticles with small-diameter, small-size distribution and with a low agglomeration level (Herlin-Boime et al. 2004). Such particles are used, for example, as precursors for surface treatments of ceramics. A similar technique, called laser ablation, is also used for the production of CNTs and nanowires.

### 1.2.5 Electrochemical Deposition/Electroplating

Electrochemical deposition is a technique used for manufacturing ordered arrays of nanomaterials, like quantum dots on a flat surface, for coating different surfaces or for patterning thin films (Allred et al. 2005). It represents one of the most powerful techniques that allow obtaining high-density and high-aspect ratio designs, with excellent reproducibility of the process and with great precision of the final products. Similarly to lithographic techniques, electrochemical deposition requires a polymer mask through which metal is deposited. Differently from other mask-based techniques like isotropic etching, ion milling, and RIE, electrochemical deposition avoids the problem of shadowing near edges or interfaces and builds structures atom by atom. The techniques rely on the presence of an electric field in a solution, which allows metal ions to discharge and conform to the smallest features of a mold.

Metal deposition can be carried on under two major techniques (Madou 2002): with current (pulsed or DC) or electroless (via catalytic, exchange, or electrophoretic reactions). The first method takes place in an electrolytic cell and involves a reaction under an imposed bias and current flow. In this case, important process parameters to control are pH, current density, temperature, agitation, and solution composition. The second technique, the electroless deposition, is based on a substantial oxidation reaction that replaces the dissolution of a sacrificial substrate.

### 1.2.6 Spraying Synthesis

Spraying processes are a viable technology for coating large area devices, and represent a simple and inexpensive alternative to produce sensing devices and nano-sized coating of surfaces (Mooney and Radding 1982). Spraying processes take place in a chamber containing a high-energy flame produced, for example, using plasma spraying equipment or using a carbon dioxide laser. A flow of reactants (gas or liquid in form of aerosols or a mixture of both) is forced into the flame and decomposes, forming particles by homogeneous nucleation and growth. The subsequent rapid cooling of the material completes the formation of nanoscale particles. In its simplest form, the reagent is simply dissolved in a carrier liquid and sprayed on a hot surface in the form of tiny droplet of ~100-nm diameter. In this case, the spray is formed from a liquid pressurized by compressed air or mechanically compressed through a small nozzle.

### 1.2.7 Atomic or Molecular Condensation

Condensation is a well-known technique used to produce primarily metal-based nanoparticles. The process takes place in a vacuum chamber: first, a bulk piece of metallic material is heated, melted, and vaporized to produce a stream of atomized matter. A gas flow (either inert or reactive) is then introduced into the reaction chamber to rapidly cool the metallic vapor. The cooled metallic atoms form condensed liquid nanoparticles, which then coalesce in a controlled environment, preserving a spherical shape with smooth surfaces. As the liquid particles are further cooled, they solidify and stop growing. The so-formed nanoparticles are very reactive and sensitive to agglomeration, which is prevented by adding coating or surfactants to keep them separated. If oxygen is present in the gaseous stream of the second chamber, the reaction results in the formation of metal oxide nanoparticles.

### 1.2.8 Supercritical Fluid Synthesis

Supercritical fluids are materials forced in a state above their critical point by regulating their pressure and temperature. They are obtained by vaporizing a liquid in a closed chamber, until the vapor phase becomes as dense as the liquid phase.

This supercritical state is very sensitive to the process' parameters and tunable. Methods using fluids in such a unique state are now used for the synthesis of nanoparticles, exploiting the rapid expansion of the supercritical solution, and they are also used for coating and filling nanostructures (Ye et al. 2004). The most common reaction media used in this process are supercritical carbon dioxide ($scCO_2$) and water ($scH_2O$) because they are nontoxic, nonflammable, inexpensive, and environmentally benign. In addition, they ensure much higher reaction rates and an increased selectivity. Although it has many advantages, supercritical fluid synthesis remains technically demanding, and it is still being adapted for large-scale production.

## 2 Conclusions

Top-down and bottom-up approaches are complementary in nanofabrication and comprise a numerous set of techniques. Their use and development have been at the center of material research for the past few decades, promising a wide range of applications in (non)biologically related nanotechnology fields. The use of a particular approach is often dictated by the specific need for a particular characteristic, shape, or composition of the nanomaterials for the different applications. Often, methods are used in combination to complement each other.

This chapter presented a brief overview of the most common techniques used for the fabrication of nanostructured bulk materials and nanomaterials with potential applications in biology and bioengineering fields. Most of the techniques described are continuously evolving even as we write and they represent a wide and active area of research in physics, chemistry, and engineering fields. A deep understanding of the mechanisms behind all the nanofabrication methods requires a broad and profound knowledge of the interdisciplinary aspects that regulate the properties of the materials and of the processes used. The details of the chemistry, physics, and mechanics of these methods go beyond the scope of this chapter and can be found in the related technical literature.

## References

Adams A (1988) In VLSI Technology. New York: McGraw-Hill.

Akerman M (2002) Nanocrystal targeting in vivo. Proc Natl Acad Sci 99:12617–12621.

Allred D, Sarikaya M, Baneyx F, Schwartz D (2005) Electrochemical Nanofabrication Using Crystalline Protein Masks. Nano Letters 5,4,:609–613.

Alwitt RS (2002) Electrochemistry Encyclopedia. In. Cleveland Ohio (http://electrochem.cwru.edu/encycl/).

Asoh H NK, Nakao M, Tamamura T and Masuda H (2001) Conditions for Fabrication of Ideally Ordered Anodic Porous Alumina Using Pretextured Al. Journal of the Electrochemical Society 148:B152-B156.

Brodie I, Muray, J J (1982) The Physics of Microfabrication, Plenum Press, New York.

Chan W, Nie S (1998) Quantum dot bioconjugates for ultrasensitive nonisotopic detection. Science 281:2016–2018.

Feynman R (1959) Plenty of Room at the Bottom. In. Pasadena: Caltech.

Guizard C, Julbe A, Larbot A, Cot L (1992) Nanostructures in sol-gel derived materials: application to the elaboration of nanofiltration membranes. Journal of Alloys and Compounds 188:8–13.

Herlin-Boime N, Mayne-L'Hermite M, Reynaud C (2004) Synthesis of covalent nanoparticles by CO2 laser. Encyclopedia of Nanoscience and Nanotechnology 10:301–326.

Iijima S (1991) Helical Microtubules of Graphitic Carbon. Nature 354:56–58.

Jensen K, ed (1989) Chemical Vapor Deposition, in Microelectronics Processing: Chemical Engineering Aspects. Washington DC: American Chemical Society.

Kroto H, Heath J, O'Brien S, Curl R, Smalley R (1985) C60: Buckminsterfullerene. Nature 318:162–163.

Lakshmi BB, Dorhout PK, Martin CR (1997) Sol-Gel Template Synthesis of Semiconductor Nanostructures. Chemistry of Materials 9, 3:857–862.

Lambrechts M SW (1992) Biosensors: Microelectrochemical Devices. Phildelphia: The Institute of Physics Publishing.

Lehmann H (1991) Plasma Assisted Etching, in Thin Film Processes II. Boston: Academic Press.

Luther W (2004) System model for nanocomposites produced by sol-gel. Published by Future Technologies Division of VDI Technologiezentrum GmbH, Dusseldorf, Germany.

Madou M (2002) Fundamentals of Microfabrication: the science of miniaturization. BocaRaton: CRC Press.

Mao C, et al. (2004) Virus-Based Toolkit for the Directed Synthesis of Magnetic and Semiconducting Nanowires, Science, Vol. 303 no. 5655 pp. 213–217.

Mooney J, Radding S (1982) Spray Pyrolysis Processing. Ann Rev Mat Science 12:81–101.

Oh S, Finones R, Daraio C, Chen L-H, Jin S (2005) Growth of Nano-scale Hydroxyapatite using chemically treated Titanium oxide nanotubes. Biomater 26:4938–4943.

Piner, R.D. et al, (1999) "Dip pen" nanolythography, Science 283, 661.

Salata O (2004) Applications of Nanoparticles in Biology and Medicine. Journal of Nanobiotechnology 2:3.

Sarikaya M, Tamerler C, Jen A-Y, Schulten K, Baneyx F (2003) Molecular Biomimetics: Nanotechnology through Biology. Nature Materials 2:577–585.

Sun S, Murray C, Weller D, Folks L, Moser A (2000) Monodisperse FePt Nanoparticles and Ferromagnetic FePt Nanocrystal superlattices. Science 287:1989–1992.

Ye X-R, Lin Y, Wang C, Engelhard MH, Wang Y, Wai CM (2004) Supercritical fluid synthesis and characterization of catalytic metal nanoparticles on carbon nanotubes. Journal of Materials of Chemistry 14:908–913.

# Characterization of Nanoscale Biological Systems: Multimodal Atomic Force Microscopy for Nanoimaging, Nanomechanics, and Biomolecular Interactions

**Arjan P. Quist and Ratnesh Lal**

## 1 Introduction

Complexity in biological systems requires coordinated research efforts using techniques and approaches that are amenable to multiscale (from nano to micro and beyond) and multidimensional (including their structure, activity, and function) microsystems (e.g., cell membrane, cell organelles, biomacromolecules, and their individual as well as integrated functioning). Major experimental designs and discoveries in biological sciences have usually been preceded by major discoveries in physicochemical sciences and engineering. Our understanding of biological systems at the time and length scale of micrometer and above is reasonable. Scaling down those systems at nanometer level is challenging and mostly unexplored as yet. The classic correspondence theory of nanoscale and microscale phenomena in physical systems is not likely to be valid for biological systems given the layers of complexity in the biological systems. Hence, one needs to examine biological systems at nanoscale in both structural and temporal domains.

A major limitation of the existing techniques to obtain nanoscale structural information about biological systems, which are all hydrated by nature, is that these techniques are unsuitable for the study of hydrated systems. The second hurdle of the existing techniques is that they lack precision and resolution for interfaces, like biological membranes, subcellular organelles, connected tissues, and biomacromolecules.

Traditional tools used to obtain structural information, such as electron microscopy, X-ray diffraction, nuclear magnetic resonance (NMR), and infrared spectros-

A.P. Quist
Richmond Chemical Corporation, Oak Brook, IL, USA

R. Lal (✉)
Departments of Mechanical and Aerospace, and Bio-engineering,
University of California San Diego, La Jolla, CA, USA
e-mail: rlal@ucsd.edu

G.A. Silva and V. Parpura (eds.), *Nanotechnology for Biology and Medicine: At the Building Block Level*, Fundamental Biomedical Technologies,
DOI 10.1007/978-0-387-31296-5_3, 

copy, provide little information regarding the surface of, for example, biomolecules or living cells. Atomic force microscopy (AFM) was developed during mid 1980s (Binnig et al. 1986). Shortly after its invention, imaging in liquids was demonstrated (Drake et al. 1989), showing the potential of the instrument for biological applications. The power of AFM comes with the fact that it can combine other imaging modalities as well as physical/chemical measurements to examine both structure and function of biological materials. The second benefit of AFM is that it can routinely be operated in aqueous and physiologically relevant solutions.

## 2 Atomic Force Microscopy

### *2.1 Operating Principle*

AFM technology is based on raster scanning a cantilevered tip in the *x*–*y* plane over a sample surface, either by attaching the sample or the cantilevered tip itself to a scanner. The *z* (vertical) position of the sample (or tip) is monitored simultaneously. The scanner consists of piezoelectric ceramic elements, most often used in a tube-like arrangement, giving the ability to have a hardware resolution of 0.1 nm laterally and 0.01 nm vertically. This is achieved by making use of the property of piezoelectric materials to change size proportional to an applied electric field (Fig. 1).

The tube scanner uses a thin-walled, radially polarized piezoelectric ceramic with one inner electrode and segmented outer electrodes that can provide motion in

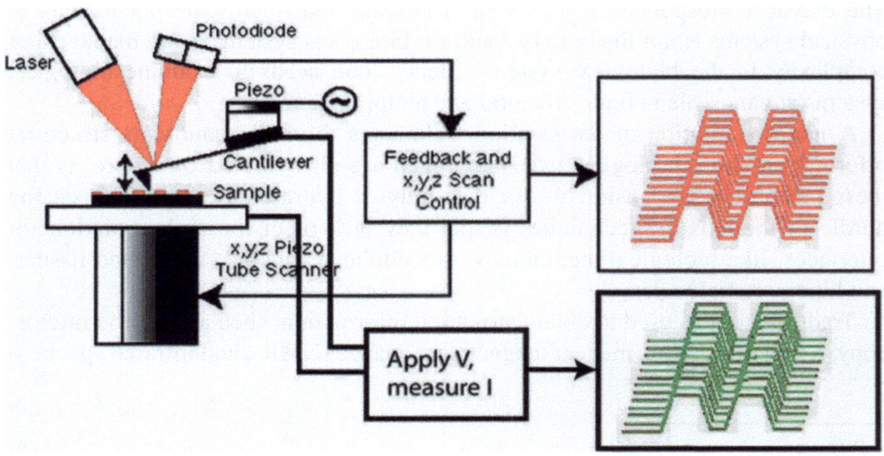


**Fig. 1** Operating principle of AFM. A cantilevered tip is scanned over the sample, maintaining constant force using feedback control to keep cantilever deflection of amplitude (measured using an optical beam deflection system) constant (*top graph*). Topography is recorded, and simultaneous additional channels can show phase, deflection, amplitude, or current flow between tip and sample (*bottom graph*)

three dimensions. In the most commonly used configuration, the deflection of the cantilever is monitored by bouncing a laser beam of the back of the cantilever, which is converted into electrical signals by a set of photodetectors (Fig. 1). The deflection of the cantilever shifts the laser reflection over the photodetectors, and a feedback electronics circuit is used to maintain a constant deflection to ensure a constant force interaction between tip and sample. The amount of $z$ variation needed to maintain the interaction force constant is plotted versus the $x$ and $y$ coordinates, hence producing a topographic image, although images may not be purely topographical (see Sect. 2.5 below). Two main modes of operation for imaging are a continuous contact mode, in which the deflection of the cantilever is maintained constant while scanning the tip over the sample, and a vibration mode in which the amplitude of vibration of an oscillating cantilever is maintained constant during scanning. In tapping mode, the phase lag between the driving circuit and the actual tip vibration can also be measured. The phase difference is usually influenced by the environmental conditions and hence provides information regarding chemical/physical properties of the sample.

## 2.2 *Spatial and Temporal Resolution*

Generally, as indicated above, the instrument hardware resolution based on the sensitivity of the piezoelectric scanner is as good as 0.1 nm laterally and 0.01 nm vertically. Of more relevance to the resolution obtained are the sample properties, especially when biological material is involved. Most often, one defines the resolution as the minimum spacing between two objects that can still be distinguished as two separate features. Thus, resolution depends mainly on the size and shape of the probing tip, as well as the material properties of the sample under investigation. On crystalline samples, images with atomic and sometimes subatomic features have been achieved (Giessibl et al. 2000) while structures less than 1 nm apart are resolved on noncrystalline arrays (Ohnesorge and Binnig 1993). On reconstituted ion channels, AFM can resolve channel substructure and resolve the subunits within each channel, distinguishing, for instance, between hexameric and pentameric connexons, and several amyloid ion channels ranging from tetramers to octamers (Lin et al. 2001; Quist et al. 2005a; Thimm et al. 2005). It is clear that the obtained resolution depends on the material to be imaged and the sample preparation methods.

The temporal resolution of the AFM depends on the speed of the feedback electronics. Typically, the scan speed in tapping mode is limited to a few tens of micrometer per second, requiring several minutes to obtain a moderately sized image with 512×512 pixels (the actual metric size of such image depends on the scanner used). Higher scan speeds are desirable not only to minimize the time spent on localizing features, but especially to be able to study the dynamic behavior of events happening on the surface to be imaged. Currently, for tapping mode, active control of a cantilever's quality factor is becoming readily available. In such systems, an extra feedback loop dampens the cantilever resonance within the drive circuit. This increases feedback efficiency and scan speed (Sulchek et al. 2002). Using an actively damped z-scanner, 240×240-nm images of Myosin in a buffer solution were acquired at a frame rate of 21 per second (Kodera et al. 2005). Using such devices, the slowest component in the AFM system is the cantilever. Since the feedback bandwidth cannot exceed 1/8 of the cantilever resonance frequency (Kodera et al. 2005), high-resonance-frequency cantilevers are needed to further improve the temporal resolution.

## 2.3 Determination of Structural Identity

Identification of features observed on biological samples can be difficult. Unless the sample contains well-defined and purified biological materials, such as for instance gap junctions or reconstituted ion channel proteins and receptors (Lal et al. 1993; Lin et al. 2001; Quist et al. 2005a), it is necessary to use complementary techniques for proper identification. Using the AFM itself, force recognition imaging (Kienberger et al. 2005) can be performed by linking specific antibodies to the tip via a spacer that recognize the presence of its antigen on the surface and result in images that do not only reflect topography but also interaction strength between the AFM tip and the sample. For further direct identification of identity, the open architecture of AFM allows for integration of optical techniques for fluorescence.

## 2.4 Integration of Optical and Assay Tools

Besides the comparison with images obtained with electron microscopy or crystallographic techniques, the design of AFM allows for easy access with optical techniques. For example, simultaneous AFM and fluorescence microscopy, i.e., epifluorescence, confocal fluorescence, and total internal reflection fluorescence (TIRF) microscopy, can be performed using fluorescent markers for biochemical/immunological identification of imaged features (Fig. 2).

As an example of the power of the multimodal imaging, simultaneous fluorescence and AFM imaging were performed on neuroblastoma cells and other cell lines (Quist et al. 2000), linking changes in cell mechanical properties (using AFM) to the functionality of hemichannels (using fluorescence). Recently, a TIRF add-on for AFM systems was developed that allows for simultaneous AFM and TIRF imaging using light-emitting diode illuminators without the requirement of specialized TIRF objectives (Ramachandran et al. 2008). Besides integration of optical tools, AFM sample supports can be designed to function as in situ lab-on-chips. For instance, using porous substrates to support biological (or other) specimens, and a conducting AFM tip, imaging can be combined with ionic conductance measurements (Ionescu-Zanetti et al. 2004; Quist et al. 2007).

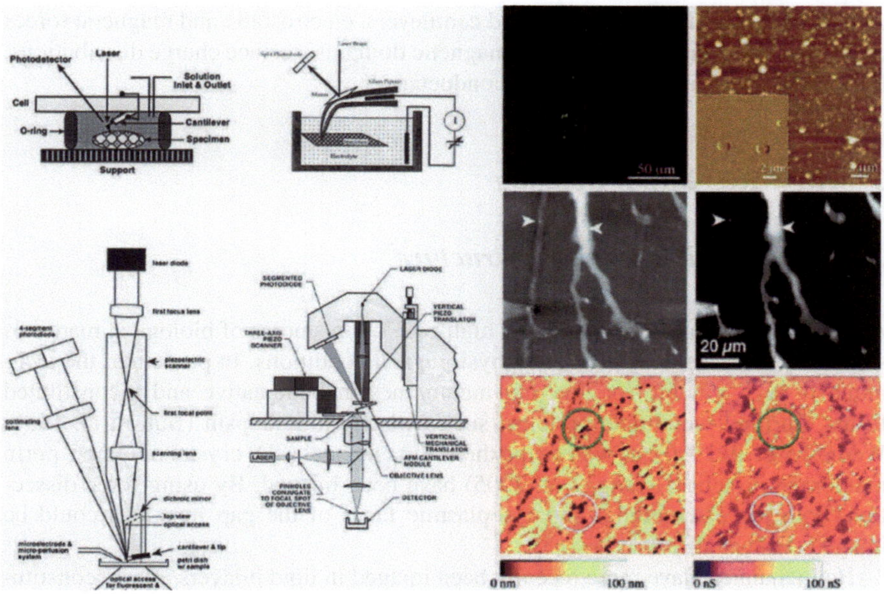

**Fig. 2** Schematic setups of multimodal AFM systems (*left*), showing (clockwise from *top left*) liquid cell, ionic conductance measurement setup, AFM integrated on optical confocal fluorescence microscope, and lens-tracking system for systems with tip attached to scanner. The *right panel* images show combined AFM and light fluorescence microscopy images of amyloid beta reconstituted vesicles (*top*), combined AFM and confocal fluorescence microscopy images of fluorescently labeled latex beads dried into a gel on a plastic diffraction grating (*middle*), and combined AFM image and electrical recording of a nucleopore synthetic membrane (*bottom*). For details, see Lal and Lin (2001)

## 2.5 Forces and Image Contrast in AFM

The deflection of the cantilever in contact mode, and the damping of vibration amplitude in tapping mode, is caused by a sum of attractive and repulsive forces. The main repulsive force is due to the overlapping of electron orbitals between tip and sample atoms. Dominant attractive force is a van der Waals interaction primarily due to nonlocalized dipole–dipole interactions (Goodman and Garcia 1991). Another strong attractive force component while imaging in air is the meniscus surface force due to adsorbed water layers. In fluids, consideration should be given to electrostatic interactions between charges and sample and tip, and structural forces, such as hydration force, solvation forces, and adhesion forces (Mechler et al. 2003; Almqvist et al. 2004; Thimm et al. 2005). Lateral forces can be measured (by torsional deflection of the cantilever) and can be quite considerable (Meyer and Amer 1990). Such forces can be used to generate images that can give valuable information on local surface chemistry differences, such as the detection of locally oxidized thiols on a topography-free surface (Pavlovic et al. 2003). In tapping mode, using conductive/magnetically coated cantilevers, electrostatic and magnetic forces can be sensed to image for instance magnetic domains, surface charge distributions, local surface capacitance, and local conductance.

# 3 High-Resolution Imaging

## 3.1 Imaging Ion Channel Structure

AFM gives a unique opportunity for high-resolution studies of biological materials since they can be examined under physiological conditions. In particular, the technique is very suitable for imaging membranes in both native and reconstituted forms. Purified membrane proteins, such as bacteriorhodopsin (Butt et al. 1990; Worcester et al. 1990), as well as the three-dimensional (3D) crystal of OmpC porin membrane protein (Hwang et al. 2005) have been imaged. By using force dissection, both the extracellular and cytoplasmic faces of the gap junctions could be imaged (Hoh et al. 1991) (Fig. 3).

Hemichannels have more recently been imaged in lipid bilayers after reconstitution in liposomes (Thimm et al. 2005). A distinct difference was observed between two populations of channel-like structures that correspond to the extracellular and cytoplasmic side of the hemichannels (Fig. 3). Furthermore, by changing calcium levels in the imaging buffer, hemichannels could be imaged both in their open (low calcium) and closed (normal calcium) states.

Using phase imaging, where the phase lag between the tip oscillation in tapping mode and the driving signal is monitored, Thimm et al. (2005) also showed surface energy measurements using AFM, observing a spike in the phase signal of open channels most likely due to exposed hydrophobic domains of the hemichannel in open state (Fig. 3). The open state of hemichannels has previously been shown to regulate cell volume and mechanics that influences tissue growth (Quist et al. 2000). Also the presence of hemichannels has been linked to induction of oxidative stress-induced apoptosis, for instance, caused by smoking (Ramachandran et al. 2007).

Other proteins that form ion channels have been reconstituted in lipid bilayers and imaged with AFM (Fig. 4) at high resolution to resolve single ion channel structures. For instance, many amyloids, including amyloid beta protein, amylin, alpha-synuclein, ADan, ABri, and Serum Amyloid A (Lin et al. 2001; Quist et al. 2005a), when reconstituted in membrane form ion channel-like structures. Resolution in such studies is sufficient to resolve a central channel and to distinguish different subunit arrangements. Electrical recording from these preparations shows that these amyloid channels are active and conduct ionic currents.

This illustrates the application of AFM in resolving the pathogenic mechanism of channelopathies. For instance, the toxicity observed upon amyloid deposition seems at least partly to be created by ion channel-induced flow of ions that disturbs the cells' homeostasis. The ability to directly image ion channels gives opportunities in testing prospective drugs and therapies on a single channel level for channelopathies.

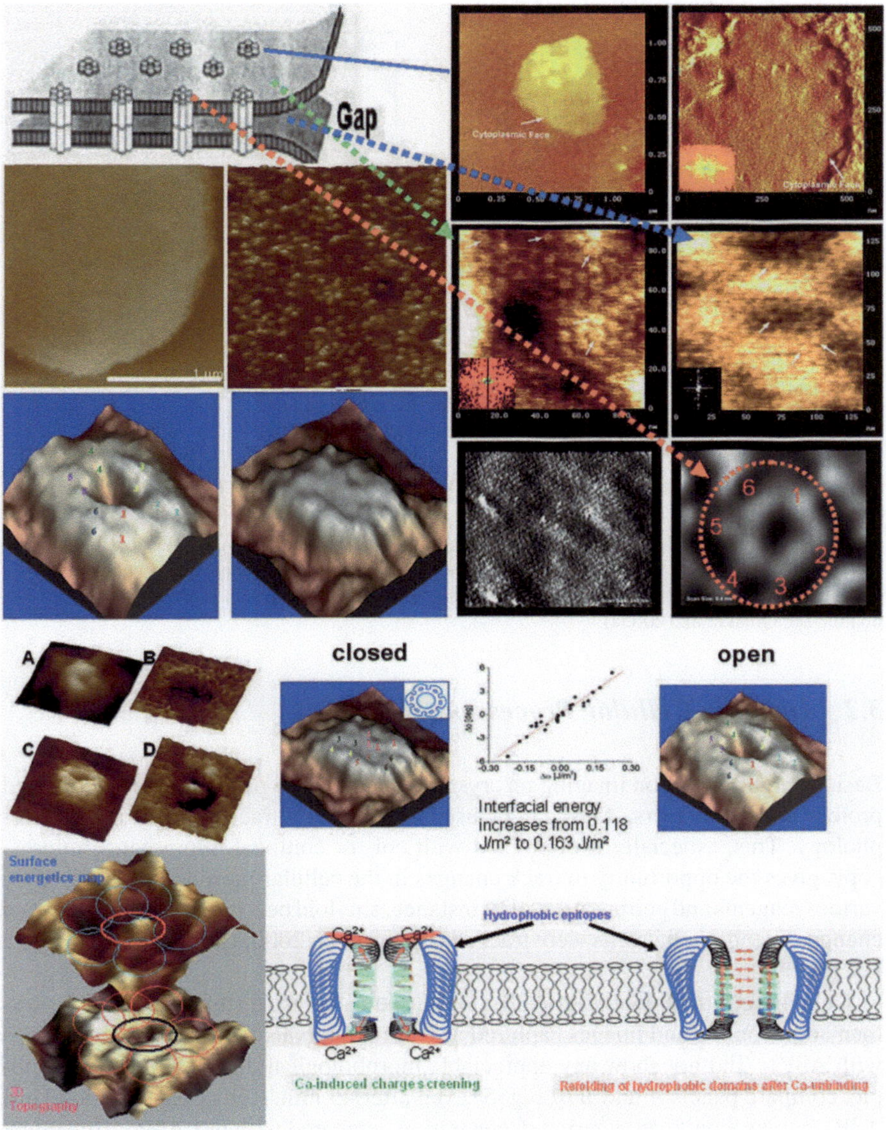


**Fig. 3** Imaging of gap junctions and hemichannels. *Top right panel* (six images) shows gap junctional plaques, imaged at cytoplasmic face and, after force dissection, extracellular face, with molecular resolution. *Top left* shows schematic and (four images) reconstituted hemichannels from bilayer to a single open and closed channel. *Bottom half* shows phase images (A–D) and energy maps (*bottom left*) of hemichannels in open and closed states. *Bottom right* section shows difference in interfacial energy of open and closed channels and a model pointing toward the role of hydrophobic interactions. For details, see John et al. (1997) and Thimm et al. (2005)

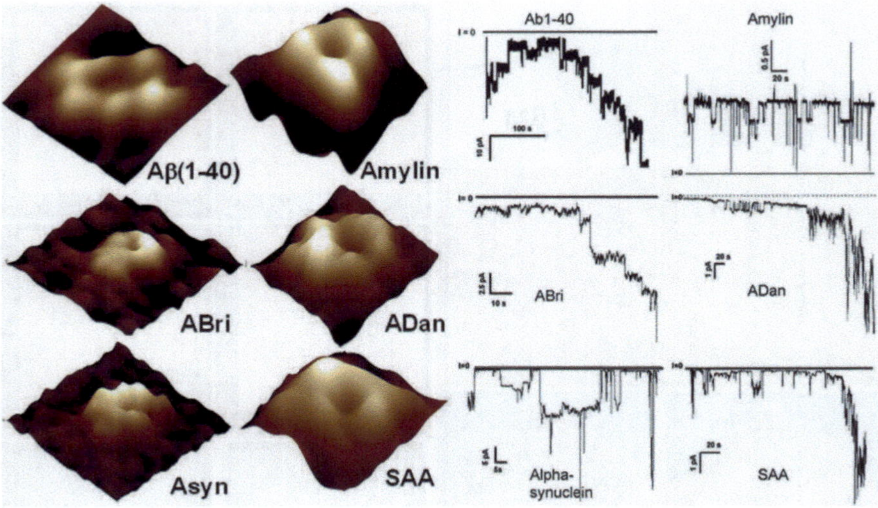


**Fig. 4** Imaging of single amyloid channels. Amyloid β (1–40), amylin, ABri, ADan, α-synuclein, and serum amyloid A (SAA) all show structures resembling ion channels after reconstitution in lipid bilayer, and show single ion channel activity as indicated by parallel electrical recording. For details, see Quist et al. (2005a)

## *3.2 Imaging Cellular Processes*

Besides high-resolution imaging on crystalline membrane patches and reconstituted protein in lipid bilayers, AFM can be used to image and track changes in cell morphology. This, especially if combined with epi- or confocal fluorescence microscopy, gives the opportunity to track changes in the cellular morphology induced by various reagents and perturbations. For instance, amyloid beta protein (1–42)-induced changes in endothelial cells were tracked (Bhatia et al. 2000; Lin et al. 2001) (Fig. 5 left panel).

Cells before treatment with $\text{AßP}_{1\text{–}42}$ are used as controls (time = 0). $\text{AßP}_{1\text{–}42}$ was then added online and images captured at 10-min intervals for 1 h. Cells not treated with $\text{AßP}_{1\text{–}42}$ did not show any appreciable morphological degeneration (for example, compare panels *A* and *B of* Fig. 5) even after 45 min. Cells treated with 50 nM $\text{AßP}_{1\text{–}42}$ show significant neuritic degeneration compared with pre-$\text{AßP}_{1\text{–}42}$ treatment (compare regions denoted by arrows in panels *C*, *D* of Fig. 5). Cells treated with 500 nM $\text{AßP}_{1\text{–}42}$ show significant changes in neuronal processes with neurite beading and loss of cell–cell contact compared with pre-$\text{AßP}_{1\text{–}42}$ treatment (compare regions denoted by arrows in panels *E*, *F* of Fig. 5). Cells treated with 5 μM $\text{AßP}_{1\text{–}42}$ show significant changes in cellular morphology with loss of neuronal processes and fragmented neurites compared with pre-$\text{AßP}_{1\text{–}42}$ treatment (compare regions denoted by arrows in panels *G*, *H* of Fig. 5). The changes in cellular morphology can be observed within 15–25 min of adding $\text{AßP}_{1\text{–}42}$.

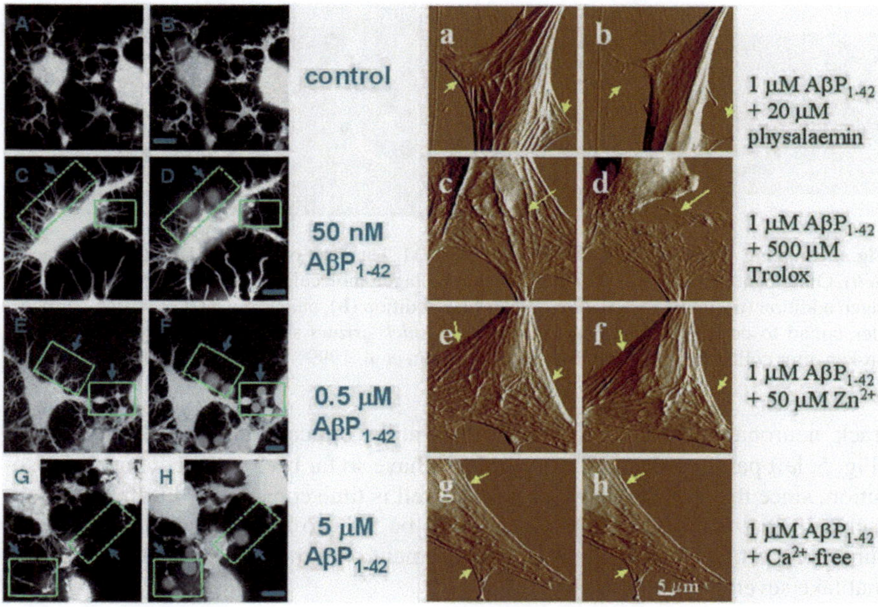


**Fig. 5** Use of the vital dye calcein and fluorescence to track neuronal death and arborization caused by addition of AβP to the buffer solution (*left*). Images taken at time 0 (control) (**A, C, E, G**) and 45 min after addition (**B, D, F, H**). Simultaneously, AFM can be used to image the cytoskeletal structure of cells to track the effects of AβP on the cells and to test various blocking agents for their performance in blocking toxicity effects. The *right panels* show that AßP-induced changes in cellular morphology were blocked by $Zn^{2+}$ and by the removal of extracellular calcium ions, but not by tachykinin (physalaemin) and antioxidants. Images (**a, c, e, g**) are controls at time 0, images (**b, d, f, h**) are taken 30 min after adding reagents. For details, see Lin et al. (2001) and Bhatia et al. (2000)

Furthermore, since the AFM operates under biological liquids, online additions of reagents or exchange of buffer conditions can be performed during AFM imaging. In this way, the blocking capability of, for instance, zinc ions and extracellular calcium ion levels to prevent such cellular morphology changes can be shown in real time (Fig. 5, right panels). Amyloid-induced morphology changes in endothelial cells were inhibited by zinc ions and by the removal of calcium ions from the medium, but not by adding tachykinin (physalaemin), antioxidants, and cadmium ions. Figure 5 (right) shows images at time 0 and 30 min after addition of the respective reagents. Another application of AFM imaging of whole live cells is the uptake of small vesicles by endocytosis into the cell. For instance, vesicles loaded with cisplatin anticancer agent were observed by both AFM and combined fluorescence to be taken up into the cell, after which live/dead assays showed the functional activity of the drug (Ramachandran et al. 2006). A multimodal AFM combined with fluorescence can give valuable additional information and conformation of processes observed with the AFM. For example, the vital dye calcein can be used to

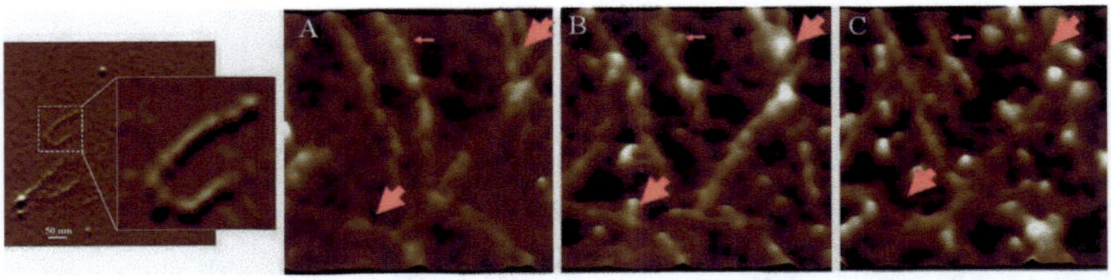


**Fig. 6** Imaging of triple-helical collagen I with AFM. The inset magnifies one collagen molecule (*left*). Other panels show collagenase binding to collagen molecules. Images are taken before collagen addition (**a**), immediately after collagenase addition (**b**), and 4 min later (**c**). Globular particles bound to collagen are indicated by *arrows*; *thick arrows* show sites, where collagen was broken after collagenase binding. For details, see Lin et al. 1999

track neuronal death and neuritic arborization optically during AFM imaging (Fig. 5, left panel). Such optical techniques have so far been a better temporal resolution, since the AFM imaging of a whole cell is time consuming. Further developments in fast-scanning AFM systems will be able to study such processes with subsecond time resolution, a large improvement over the conventional AFM scans that take several minutes to collect.

## 3.3 Single Molecule Imaging and Manipulation

Since single molecules can be imaged routinely in AFM, it is feasible to use the AFM to image real-time processes of molecules' interaction with each other. For instance, proteolysis of collagen was observed using AFM (Fig. 6) (Lin et al. 1999).

Such processes are typically monitored using SDS-PAGE which gives little information about the real-time activity and the 3D structural information of the binding of the enzyme to the molecules. Single collagenase binding events to collagen could be imaged; digestion of collagen could be followed.

With the possibility of monitoring single molecules comes the advantage of being able to manipulate the molecules as well. By tethering antibodies to the AFM tip, and after bringing the tip in contact with the sample of interest, the antigen in the sample can not only be detected, but also unfolded. For instance, bacteriorhodopsin molecules could be extracted from purple membrane and unraveled by pulling with the AFM tip (Kienberger et al. 2005).

## 3.4 Mapping Polymer Electrical Activity and Structure

There has been a large commercial and scientific interest in conjugated polymer blends. Such polymer systems have a wide range of applications ranging from microactuators (Jager et al. 2000) to chemical sensors (Hagleitner et al. 2001) and

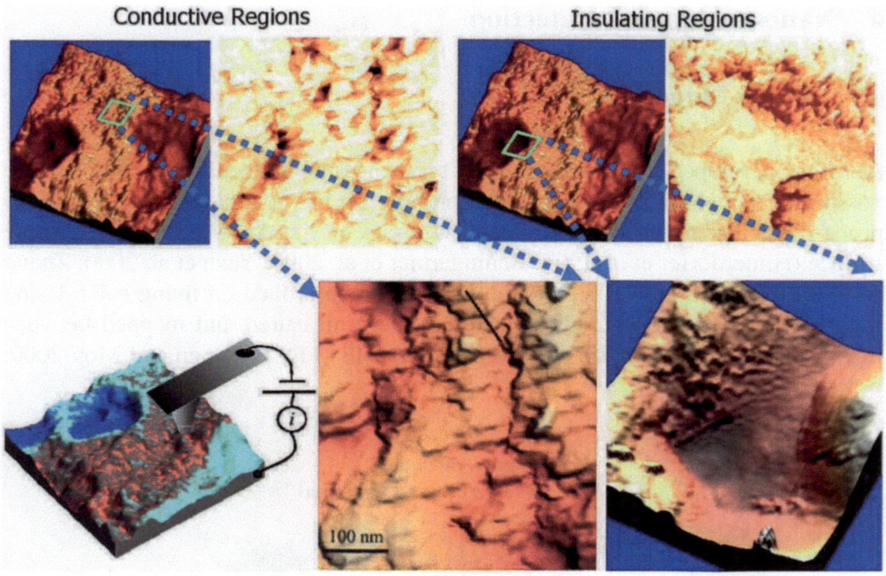


**Fig. 7** Imaging topography and conductance of polymer blends. *Top panels* show 3D overlay of conductance and topography (*left panel*), corresponding phase image (*second panel from left*) for conductive and insulating (*right panels*) regions of a thin PEDOT–PSS film. Highly conductive regions are colocalized with high roughness and high phase variability. *Bottom right* shows high-resolution images of selected regions showing detailed lamellar structure of conducting region (*bottom middle*) and nonlamellar structure of nonconducting region (*bottom right*). Schematic drawing of setup shown in *bottom left*. For details, see Ionescu-Zanetti et al. (2004)

light-emitting diodes (Carter et al. 1997). Although macroscopic structural studies are available and performed mainly by X-ray diffraction (Aasmundtveit et al. 1999), information about the effect of local morphology of polymer blends on the charge injection at the surface has been lacking. AFM can combine imaging with spectroscopic capabilities and can be used to map the local charge transfer properties in correlation to the molecular superstructure of the polymer blend (Ionescu-Zanetti et al. 2004).

Structure–function analysis of poly(3,4-ethylenedioxythiophene) doped with poly(styrene sulfonate) (PEDOT–PSS) was performed: current between the tip and the sample was measured by biasing the AFM tip while simultaneously imaging the 3D structure (Fig. 7). This multimodal AFM experiment shows that the PEDOT–PSS consists of molecular lamellae of PEDOT and PSS, with ~3-nm interlamellar distance. In places where the edges of the PEDOT lamellae are exposed to the surface and can be contacted by the AFM tip, increased current is observed and a more efficient charge injection occurs mainly along these lamellae. Phase imaging simultaneously reveals a larger variation along the regions with higher conductivity.

# 4 Nanosensing and Detection

## 4.1 Force Mapping

AFM has developed into a valuable tool for measuring molecular interactions. The possibility of linking molecules to the probing tip provides a range of options to probe molecular interactions between the probe on the AFM tip and the sample surface (Hinterdorfer et al. 1996; Baumgartner et al. 2000; Yuan et al. 2000; Zhang et al. 2002). In this way, membrane proteins can be probed on living cells. Using such modified cantilevers, adhesion forces were measured and mapped between ligands and receptors on the surface of living cells (Fig. 8) (Chen and Moy 2000;

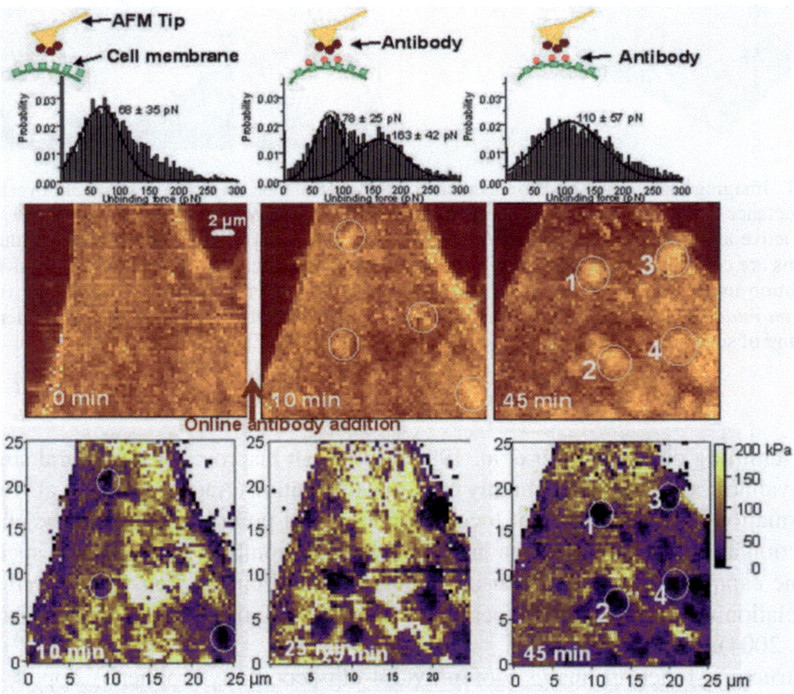


**Fig. 8** Simultaneous acquisition of topography and elasticity images before and 10 and 45 min after adding VEGF antibody (*top half*). Clustering of receptors can be observed both in topography and elasticity maps (spots labeled 1–4). *Top histograms* show distribution of adhesion forces. *Bottom half* (*left*); adhesion forces between VEGF receptor and antibody (**a**), with distribution of unbinding forces (**b**). Blocking peptide prevents the interaction (**c**). (**d**) Tapping mode image of VEGF receptors on mica (**e**) and (**f**) show adhesion forces between VEGF receptor on endothelial cell and antibody on tip without (**e**) and with (**f**) the presence of blocking peptide. *Bottom half* (*right*); AFM images of endothelial cells before (**a**) and after (**b**) addition of VEGF, showing cytoskeletal reorganization. Fluorescence images showing the presence of Flk-1, a receptor for VGEF (**c, d**) and control (**e**) with nonspecific antibody. For details, see Almqvist et al. (2004)

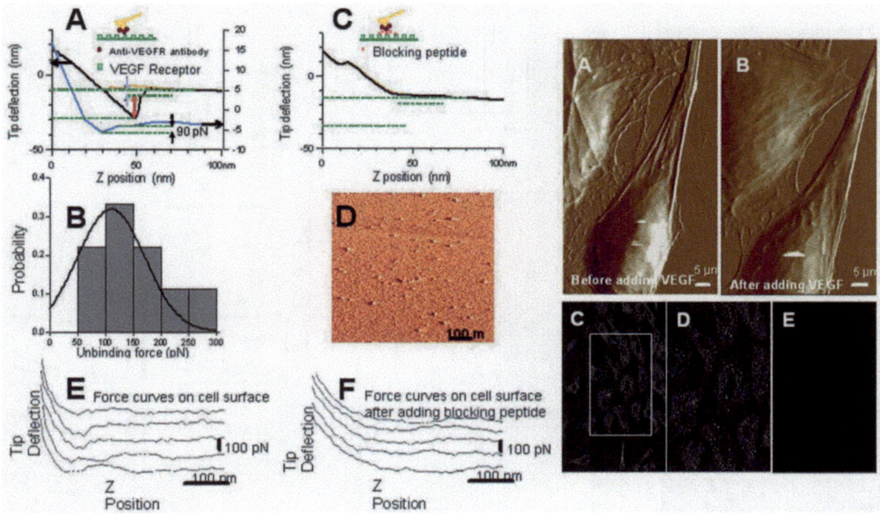


**Fig. 8** (continued)

Horton et al. 2002). Using force volume imaging (Quist et al. 2000), regional distribution as well as ligand- or antibody-induced clustering of vascular endothelial growth factor (VEGF) receptors are reported (Fig. 8) (Almqvist et al. 2004). Unbinding forces of 60–70 pN were observed, which could be reduced by competitive inhibition using antibodies.

Force mapping can also be useful for volume measurement on spherical cells and/or cells weakly attached to the substrate surface. Using combined AFM with fluorescence microscopy, Quist et al. (2000) have shown volume regulation of several cell lines through hemichannels. The added benefit of using force mapping is that not only is data obtained with respect to unbinding forces, but simultaneous with imaging one can map the stiffness of the cell membrane. The stiffness is obtained by looking at the approach force distance curves in each pixel and results in an elasticity map. Almqvist et al. (2004) showed clustering of receptors using this technique, and Quist et al. (2000) showed the cytoskeletal stiffness change induced by volume change in the cell. By using antibodies conjugated on the AFM tip by linker molecules that act on different regions of the connexin peptide that makes up the hemichannels, epitopes that open and close the channels can be mapped in response to variations in extracellular calcium concentration (Liu et al. 2006). Such mapping or channel domains and how they react to different antibodies and environmental distraction are a great help in designing drugs for ion channel-related diseases.

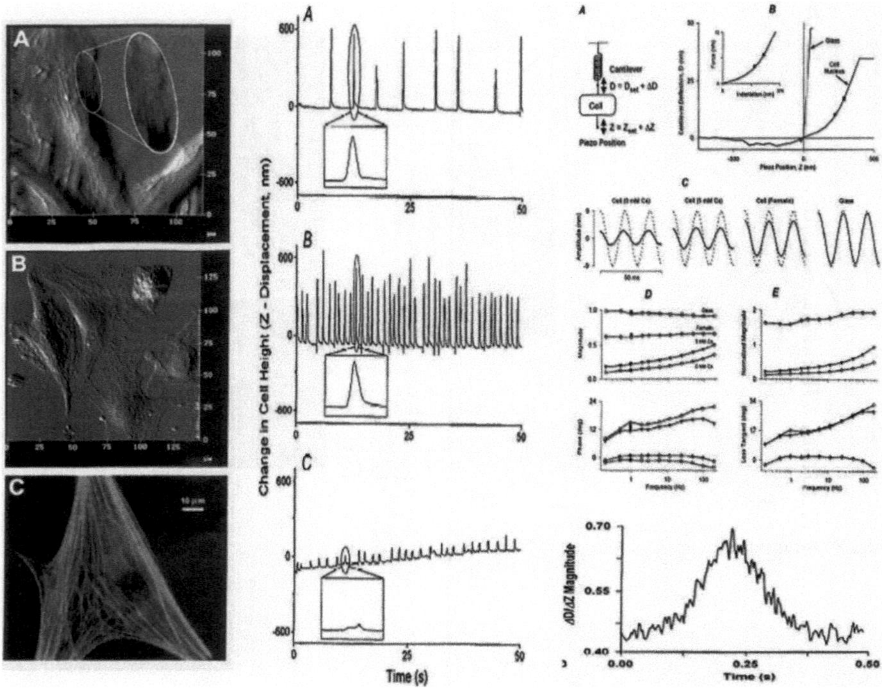

**Fig. 9** *Left panels*: AFM image of live myocytes (**a**, inset shows myofibril structure), deflection image of fixed cell (**b**), and fluorescence image after staining for F actin. *Center panels* (*vertical*) show local contractile activity of beating atrial cell recorded with AFM tip, under 1.8 mM (**a**) and 5 mM (**b**) extracellular calcium ions, and 4 mM butanedione monoxime with 5 mM calcium concentration (**c**). *Right panels* (*vertical*): Micromechanical properties of quiescent atrial cells. Model (**a**) and indentation data (**b**). Response of a cell (*solid line*) to sinusoidal *Z* perturbation (*dashed line*) (**c**, four experimental conditions). Frequency spectra of magnitude and phase of transfer function (**d**) on cells, and (**e**) using mechanical analogue. *Bottom right*: Changes in cell stiffness during single contraction. For details, see text and Shroff et al. (1995)

## 4.2 Cell Mechanics

Proper cellular function is dependent on biochemical processes as well as cytoskeletal structure. Reorganization of cytoskeleton and changes in its mechanical properties play key roles in cell development and growth (Lal et al. 1995). The local mechanical properties of a cell are closely associated with biochemical gradients across the membrane, but most techniques to study single cells average over the entire cell (Brady 1991). AFM gives the opportunity to measure the mechanical properties of cells with high spatial resolution. For instance, rat atrial myocytes were imaged showing clearly the cytoskeletal network beneath the cell membrane and myofibrillar structure [Fig. 9, left panel (Shroff et al. 1995)].

Using a constant cantilever deflection maintained by feedback, contractile activity and the change in contractile activity using perturbations in the buffer environment (Fig. 9, middle panel) can be interrogated while cell stiffness was examined (Fig. 9, right panel) using a sinusoidal scanner movement. Different cantilever deflections were observed under different buffer conditions for the same imaging force. This demonstrates the possibility to quantify the coupling between subcellular substrates to cellular functions, such as contraction, migration, growth, and differentiation.

### *4.3 Nanoflow and Viscosity*

Liquid viscosity is hard to measure with high precision and in small volumes. Traditionally, ultrasonic devices are used to measure viscosity (Jensenius et al. 2000). They operate at MHz frequency at which the measured viscosity of non-Newtonian fluids can differ from values obtained using low-frequency measurements (Wee et al. 2005). Flexural-mode resonance devices, such as microfabricated cantilevers, may be more reliable since they allow for measurement at lower frequencies. Using an optical detection in standard AFM equipment, viscous drag was measured using a piezoelectric actuator to vibrate an AFM silicon cantilever (Hauptmann et al. 1998). Other ways of using AFM to measure liquid viscosity include measuring the torsion in an AFM cantilever while scanning a whisker tip inside the liquid (Mechler et al. 2004).

The last technique allows viscosity measurement in a thin fluid film requiring extremely low volumes. The lateral force imaging technique is used, in which the cantilever is scanned sideways and the torsion in the lever due to friction is measured (Fig. 10). From the torsion, the drag forces can be determined as a function of the penetration depth of the AFM tip in the liquid. Theoretical calculation of the drag coefficient corresponds well with the experimental values obtained for water–glycerol mixtures. With the increased use of microfluidics in screening platforms and parallel assays and the ability to integrate microcantilevers into the fluidic chip could be used not only to measure viscosity of sample fluids, but also to assess the presence of various proteins/antibodies in the sample.

## 5 AFM in Biomaterials

### *5.1 Nanoelasticity and Nanopatterning in Biomaterials*

Biological fibers have nanomechanical properties that depend on their morphology as well as the chemical heterogeneity of their constituent subunits. The correlation between the mechanical properties and the heterogeneous subunits on a nanoscale has been limited. Using AFM, such correlation studies are possible (Parbhu et al. 1999).

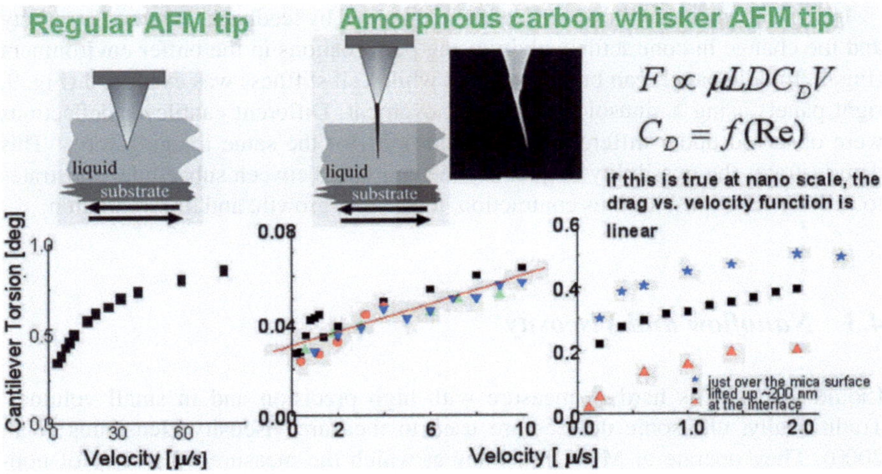


**Fig. 10** Schematic of viscosity measurement using a regular or carbon whisker AFM tip. The *right* picture shows electron microscope image of a whisker tip. *Bottom graphs* show cantilever torsion while scanning in the lateral force mode. For details, see Mechler et al. (2004)

AFM has been able at a small scan size to show the features of a cross-sectioned wool fiber with similar amount of detail as transmission electron microscopy (TEM) (Fig. 11, left panels). To study the mechanical properties of the different constituents of the fibers, AFM force mapping has been used. The force curves gives information about the stiffness and elastic modulus of the sample regions. For instance, it was shown that the exocuticle part of wool fibers has the highest elastic modulus (19.8 GPa) while the endocuticle and cortical regions of the fiber cross section have significant lower modulus (3.8 and 4.0 GPa, respectively). The indentations made by a diamond tipped cantilever, all made with the same indentation force, look indeed distinctly different for the different regions (Fig. 11, right panel). Furthermore, the AFM can give conclusive evidence of the role of disulfide bonds in the fiber stiffness. Reduction of such bonds, abundant in the exocuticle, using dithiothreitol resulted in a reduction of modulus to 0.3 GPa (Parbhu et al. 1999). This has shed new light on the role of the exocuticle (small in mass proportion) in the rigidity of wool fibers.

Using the unique feature of defined nanoindentation, one can create specific patterned structures, like nanocavity, nanopores, nanowells, and nanochannels. Significantly, such nanopatterned features would also provide clues about the building materials and would be helpful in advancing the field of biomimetics.

## 5.2 *Nanomechanics of Human Bone*

The mechanical properties of bone and more specifically the failure mechanisms of healthy and diseased bone are receiving large scientific as well as medical interest. The organic moiety in bone has a great influence on the mechanical properties of the

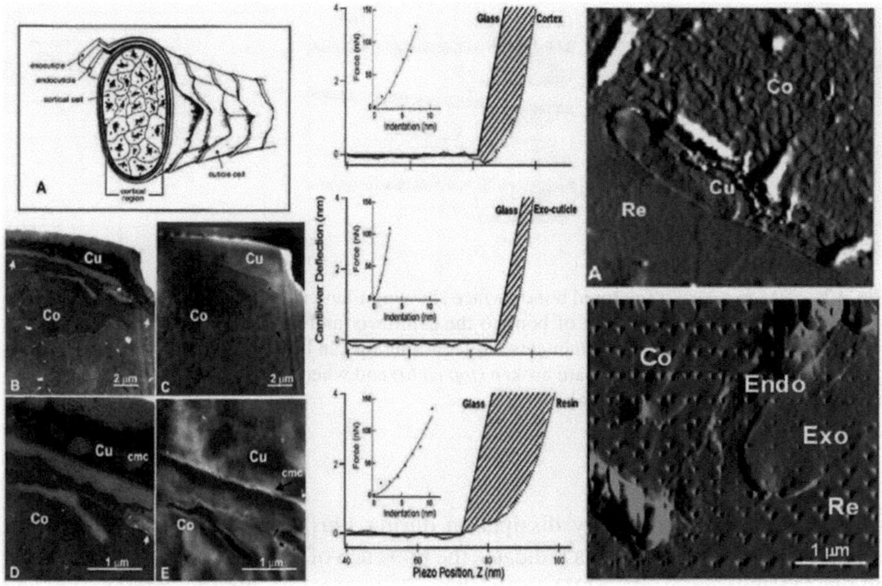


**Fig. 11** *Left*: Schematic diagram of wool fiber (**a**). TEM image [(**b**), higher magnification in (**d**)] and AFM image [(**c**), higher magnification in (**e**)] of same wool fiber region. Cu, cuticle cells and some portion of the cortical region (Co). AFM and TEM images correlate well with respect to identified subcellular structures. *Center*: Force curves on cortex (*top*), exocuticle (*middle*), and embedding resin (*bottom*). *Solid line* is collected on hard glass surface. *Larger shaded area* indicates more elastic surface. *Top right* image shows corresponding AFM image; the cortex (Co), cuticle (Cu), and embedding resin (Re). *Bottom right* image shows surface after indentation with diamond tip to create specific patterns on wool fiber; exocuticle (Exo), endocuticle (Endo), and cortex (Co). For details, see Parbhu et al. (1999)

complete biocomposite (Weiner and Wagner 1998). Degradation of this nonfibrillar organic matrix (Braidotti et al. 1997), for instance by heat treatment, leads to an alteration of the failure mechanism from delamination of the bone along the contours of the trabeculae to a more rugged, nondirectional and diffuse crack propagation (Fantner et al. 2004). Similar degradation of the organic matrix leading to brittle bones may be involved in the increased brittleness of older or osteoporotic bone. Using AFM, the mechanics of bone and the importance of the organic matrix can be studied at the nanoscale. Both scanning electron microscopy (SEM) and AFM studies show that mineralized collagen fibrils are interconnected by a glue-like organic matrix that hold the fibrils together (Fig. 12, left panel) (Fantner et al. 2005) and are stretched out upon separation of the collagen fibrils.

To test the mechanical strength of the organic matrix material, pieces of bone were attached to both the AFM tip and cantilever, and force spectroscopy resulted in pulling curves that show the filaments breaking in a step-like manner (Fig. 12, middle and right panels) (Fantner et al. 2005). Furthermore, from the force curves, the energy dissipation during separation of bone fibrils can be calculated. Since AFM is performed in buffer solution, the effect of ion concentration can also be

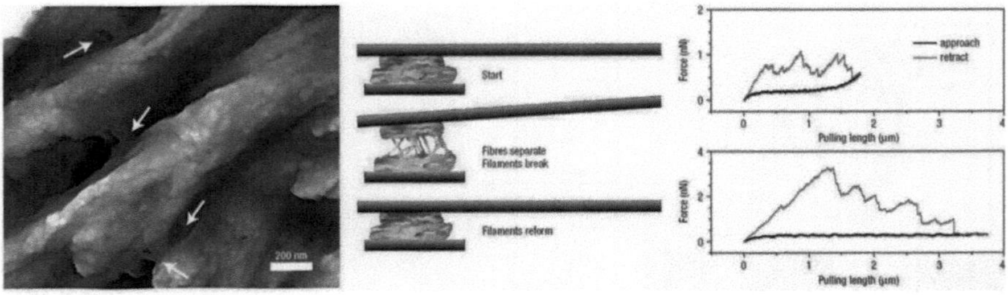


**Fig. 12** AFM image of a fractured bone surface showing filaments (*arrows*) between neighboring fibrils (*left*). By attaching a piece of bone to the cantilever and pressing it onto another piece of bone (*center*), forces exerted by filaments while pulling up can be measured. Representative pulling curve, where not all filaments are broken (*top right*) and where all filaments are broken (*bottom right*). For details, see Fantner et al. (2005)

studied, showing that energy dissipation during fibril separation is greater when calcium ions are present. This indicates the presence of calcium-mediated sacrificial bonds (Thompson et al. 2001) in the bone between two binding regions on one polymer, two polymers, or a polymer and a mineral plate.

# 6 Nanodevices for Biological Applications

## 6.1 Parallel Arrays Patch Clamping and Sensors

Patch clamp recording is one of the main techniques employed in electrophysiological studies. Generally, a precisely pulled glass pipette is brought in the vicinity of a cell of interest under an optical microscope using a micromanipulator (Sakmann and Neher 1983). This results in the technique being slow and having a low throughput, making it not very suitable for use in, for instance, proteomics and drug discovery development. This could be overcome using an automated on-chip patch clamp (Xu et al. 2001).

An example of such an on-chip microchannel planar patch clamp device is shown in Fig. 13. The middle panels show a HeLa cell being introduced in the microfluidic system and being trapped at one of the channels. Cell deformation can be observed, and current traces show a seal resistance of 144 MOhm (Seo et al. 2004). Such devices using a cell reservoir linked to patch channels allow for simultaneous optical and electrical recording which facilitates studying the role of ion channels with respect to cellular functions. Future devices can be designed with an improved patch pore geometry and surface treatment to obtain a better seal necessary for single-channel conductance studies. Since the cell reservoir is linked to many patch clamp pores, a parallel readout is possible analyzing multiple cells simultaneously.

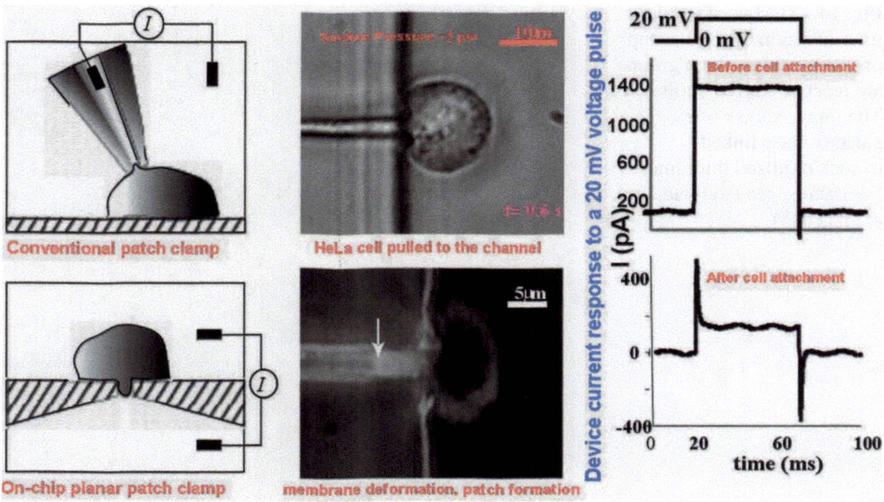


**Fig. 13** Schematic drawing of a conventional patch clamp and an on-chip patch clamp (*left*). Cells can be pulled into one of the on-chip patch pores (*top middle*), resulting in membrane deformation and patch formation (*bottom middle*). Resulting current measured before (*top right*) and after cell attachment (*bottom right*). For details, see Seo et al. (2004)

Other parallel readout techniques based on AFM include cantilever arrays that allow for analysis of large areas with high resolution, and can probe for instance multiple live cells for their elastic properties or the presence of receptors in the cell membrane. Similarly, coating a parallel array of cantilever with a different material or reagent on each lever results in a "chemical nose" that can sense a variety of chemicals or toxins in very small volumes or could read out the viscosity of sample fluid.

## 6.2 *Surface Chemical Array Modifications*

A novel technique that has been applied more and more in recent years to construct stable nanoscale patterns on surfaces is dip pen nanolithography (DPN) (Piner et al. 1999). The technique is based on the localized deposition of reagents from the AFM tip to the surface using a solvent meniscus between the tip and sample surface. The main benefit of DPN is that it is a one-step process, does not require resists, and can achieve line widths of 10–15 nm, a resolution not achieved with photolithography or microcontact printing (Quist et al. 2005b). Bonding of reagent to the sample surface is usually achieved by chemisorption through thiol chemistry or by electrostatic interactions. Using DPN, it is possible to create high-density arrays of, for instance, DNA or protein (Fan et al. 2002) or to write structures with conducting polymers on charged semiconductor surfaces (Lim and Mirkin 2002) that can be used for assay chip technology.

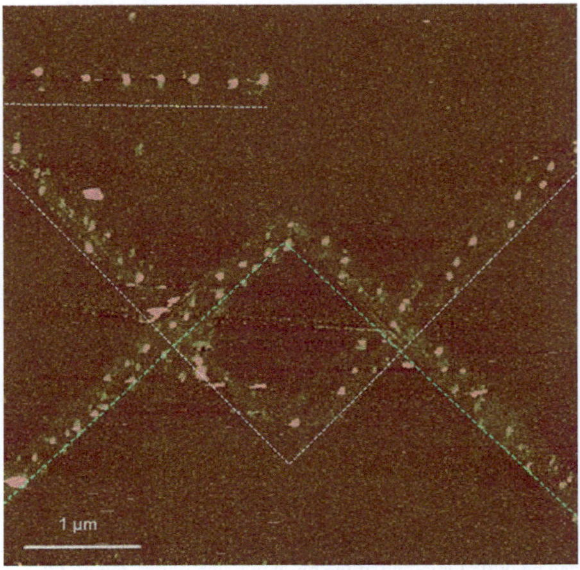


**Fig. 14** AFM used to "draw" lines of oxidized thiol groups on silicon chips. Such groups are reactive to free thiols. The image shows beta-galactosidase linked to such oxidized thiol lines. For details, see Pavlovic et al. (2003)

Another method to selectively position, for instance, protein molecules on surfaces with high resolution is to use the AFM tip to locally oxidize a surface to create a reactive group that can bond to the biomolecules of interest. For instance, 3-mercaptopropyltrimethoxysilane can be used to thiolate a silicon wafer, after which the AFM tip can be used to locally oxidize the thiol groups into thiolsulfinates/thiolsulfonates. The latter groups are reactive and covalently bind to free thiol groups on biomolecules (Fig. 14) (Pavlovic et al. 2003). Beta-galactosidase, an enzyme rich in exposed and structurally now-essential free thiol groups, was linked covalently to silicon wafers using this technique, resulting in line widths as low as 70 nm. Using such systems based on AFM technology, it would be possible to design assay chips with a much higher density of components for screening, and with a parallel readout using multicantilever systems may be developed into new high-throughput screening tools.

## 7 Outlook

The simple design and invariance to the operating environment allow AFM to be integrated with other techniques for simultaneous structure–function correlation studies. The integration of the AFM and fluorescence microscope is one such exciting development. Other combinations, specifically a combined near-field differential scanning optical microscope with AFM, near-field optical imaging, electrophysiological recordings, and the AFM-based NMR imaging, provide promising avenues for nanoscale structural and functional characterization of complex biological systems. Moreover, the flexibility and high resolution capability of these integrated tools point toward a unique window through which new and exciting information could be harnessed about and from multiscale biological systems. The AFM provides a toolbox in combination with other techniques which can not only elucidate mechanisms behind biological functioning, but also could assist in finding ways to treat disease as an aid in drug development. We should also expect to have next generation of AFMs with faster and softer imaging features that would allow us not only to understand real-time conformational changes in nanoscale biological systems, but would also allow designing and fabricating biomimetic systems.

## References


Aasmundtveit KE, Samuelsen EJ, Pettersson LAA, Inganas O, Johansson T, Feidenhans R (1999) Structure of thin films of poly(3,4-ethylenedioxythiophene). Synthetic Metals 101:561–564.

Almqvist N, Bhatia R, Primbs G, Desai N, Banerjee S, Lal R (2004) Elasticity and adhesion force mapping reveals real-time clustering of growth factor receptors and associated changes in local cellular rheological properties. Biophysical Journal 86:1753–1762.

Baumgartner W, Hinterdorfer P, Ness W, Raab A, Vestweber D, Schindler H, Drenckhahn D (2000) Cadherin interaction probed by atomic force microscopy. Proceedings Of The National Academy Of Sciences Of The United States Of America 97:4005–4010.

Bhatia R, Lin H, Lal R (2000) Fresh and globular amyloid beta protein (1–42) induces rapid cellular degeneration: evidence for A beta P channel-mediated cellular toxicity. Faseb Journal 14: 1233–1243.

Binnig G, Quate CF, Gerber C (1986) Atomic Force Microscope. Physical Review Letters 56:930–933.

Brady AJ (1991) Mechanical-Properties Of Isolated Cardiac Myocytes. Physiological Reviews 71:413–428.

Braidotti P, Branca FP, Stagni L (1997) Scanning electron microscopy of human cortical bone failure surfaces. Journal Of Biomechanics 30:155–162.

Butt HJ, Downing KH, Hansma PK (1990) Imaging The Membrane-Protein Bacteriorhodopsin With The Atomic Force Microscope. Biophysical Journal 58:1473–1480.

Carter SA, Angelopoulos M, Karg S, Brock PJ, Scott JC (1997) Polymeric anodes for improved polymer light-emitting diode performance. Applied Physics Letters 70:2067–2069.

Chen A, Moy VT (2000) Cross-linking of cell surface receptors enhances cooperativity of molecular adhesion. Biophysical Journal 78:2814–2820.

Drake B, Prater CB, Weisenhorn AL, Gould SAC, Albrecht TR, Quate CF, Cannell DS, Hansma HG, Hansma PK (1989) Imaging Crystals, Polymers, And Processes In Water With The Atomic Force Microscope. Science 243:1586–1589.

Fan YW, Cui FZ, Hou SP, Xu QY, Chen LN, Lee IS (2002) Culture of neural cells on silicon wafers with nano-scale surface topograph. J Neurosci Methods 120:17–23.

Fantner GE, Hassenkam T, Kindt JH, Weaver JC, Birkedal H, Pechenik L, Cutroni JA, Cidade GAG, Stucky GD, Morse DE, Hansma PK (2005) Sacrificial bonds and hidden length dissipate energy as mineralized fibrils separate during bone fracture. Nature Materials 4:612–616.

Fantner GE, Birkedal H, Kindt JH, Hassenkam T, Weaver JC, Cutroni JA, Bosma BL, Bawazer L, Finch MM, Cidade GAG, Morse DE, Stucky GD, Hansma PK (2004) Influence of the degradation of the organic matrix on the microscopic fracture behavior of trabecular bone. Bone 35:1013–1022.

Giessibl FJ, Hembacher S, Bielefeldt H, Mannhart J (2000) Subatomic features on the silicon (111)–(7×7) surface observed by atomic force microscopy. Science 289:422–425.

Goodman FO, Garcia N (1991) Roles Of The Attractive And Repulsive Forces In Atomic-Force Microscopy. Physical Review B 43:4728–4731.

Hagleitner C, Hierlemann A, Lange D, Kummer A, Kerness N, Brand O, Baltes H (2001) Smart single-chip gas sensor microsystem. Nature 414:293–296.

Hauptmann P, Lucklum R, Puttmer A, Henning B (1998) Ultrasonic sensors for process monitoring and chemical analysis: state-of-the-art and trends. Sensors And Actuators A-Physical 67:32–48.

Hinterdorfer P, Baumgartner W, Gruber HJ, Schilcher K, Schindler H (1996) Detection and localization of individual antibody-antigen recognition events by atomic force microscopy. Proceedings Of The National Academy Of Sciences Of The United States Of America 93:3477–3481.

Hoh JH, Lal R, John SA, Revel JP, Arnsdorf MF (1991) Atomic Force Microscopy And Dissection Of Gap-Junctions. Science 253:1405–1408.

Horton M, Charras G, Lehenkari P (2002) Analysis of ligand-receptor interactions in cells by atomic force microscopy. Journal Of Receptor And Signal Transduction Research 22:169–190.

Hwang NS, Kim MS, Sampattavanich S, Baek JH, Zhang Z, Elisseeff J (2005) The Effects of Three Dimensional Culture and Growth Factors on the Chondrogenic Differentiation of Murine Embryonic Stem Cells. Stem Cells:113–123.

Ionescu-Zanetti C, Mechler A, Carter SA, Lal R (2004) Semiconductive polymer blends: Correlating structure with transport properties at the nanoscale. Advanced Materials 16:385-+.

Jager EWH, Smela E, Inganas O (2000) Microfabricating conjugated polymer actuators. Science 290:1540–1545.

Jensenius H, Thaysen J, Rasmussen AA, Veje LH, Hansen O, Boisen A (2000) A microcantilever-based alcohol vapor sensor-application and response model. Applied Physics Letters 76:2615–2617.

John SA, Saner D, Pitts JD, Holzenburg A, Finbow ME, Lal R (1997) Atomic force microscopy of arthropod gap junctions. Journal Of Structural Biology 120:22–31.

Kienberger F, Kada G, Mueller H, Hinterdorfer P (2005) Single molecule studies of antibody-antigen interaction strength versus intra-molecular antigen stability. Journal Of Molecular Biology 347:597–606.

Kodera N, Yamashita H, Ando T (2005) Active damping of the scanner for high-speed atomic force microscopy. Review Of Scientific Instruments 76.

Lal R, Lin H (2001) Imaging molecular structure and physiological function of gap junctions and hemijunctions by multimodal atomic force microscopy. Microscopy Research And Technique 52:273–288.

Lal R, Kim H, Garavito RM, Arnsdorf MF (1993) Imaging Of Reconstituted Biological Channels At Molecular Resolution By Atomic-Force Microscopy. American Journal Of Physiology 265:C851-C856.

Lal R, Drake B, Blumberg D, Saner DR, Hansma PK, Feinstein SC (1995) Imaging Real-Time Neurite Outgrowth And Cytoskeletal Reorganization With An Atomic-Force Microscope. American Journal Of Physiology-Cell Physiology 38:C275-C285.

Lim JH, Mirkin CA (2002) Electrostatically driven dip-pen nanolithography of conducting polymers. Advanced Materials 14:1474-+.

Lin H, Clegg DO, Lal R (1999) Imaging real-time proteolysis of single collagen I molecules with an atomic force microscope. Biochemistry 38:9956–9963.

Lin H, Bhatia R, Lal R (2001) Amyloid beta protein forms ion channels: implications for Alzheimer's disease pathophysiology. Faseb Journal 15:2433–2444.

Liu F, Arce FT, Ramachandran S, Lal R (2006) Nanomechanics of Hemichannel Conformations: connexin flexibility underlying channel opening and closing. J Biol Chem 281:23207–23217.

Mechler A, Kokavecz J, Heszler P, Lal R (2003) Surface energy maps of nanostructures: Atomic force microscopy and numerical simulation study. Appl Phys Lett 82:3740–3742.

Mechler A, Piorek B, Lal R, Banerjee S (2004) Nanoscale velocity-drag force relationship in thin liquid layers measured by atomic force microscopy. Applied Physics Letters 85:3881–3883.

Meyer G, Amer NM (1990) Simultaneous Measurement Of Lateral And Normal Forces With An Optical-Beam-Deflection Atomic Force Microscope. Applied Physics Letters 57:2089–2091.

Ohnesorge F, Binnig G (1993) True Atomic-Resolution By Atomic Force Microscopy Through Repulsive And Attractive Forces. Science 260:1451–1456.

Parbhu AN, Bryson WG, Lal R (1999) Disulfide bonds in the outer layer of keratin fibers confer higher mechanical rigidity: Correlative nano-indentation and elasticity measurement with an AFM. Biochemistry 38:11755–11761.

Pavlovic E, Oscarsson S, Quist AP (2003) Nanoscale site-specific immobilization of proteins through electroactivated disulfide exchange. Nano Letters 3:779–781.

Piner RD, Zhu J, Xu F, Hong SH, Mirkin CA (1999) "Dip-pen" nanolithography. Science 283:661–663.

Quist A, Doudevski L, Lin H, Azimova R, Ng D, Frangione B, Kagan B, Ghiso J, Lal R (2005a) Amyloid ion channels: A common structural link for protein-misfolding disease. Proceedings Of The National Academy Of Sciences Of The United States Of America 102:10427–10432.

Quist AP, Pavlovic E, Oscarsson S (2005b) Recent advances in microcontact printing. Analytical And Bioanalytical Chemistry 381:591–600.

Quist AP, Rhee SK, Lin H, Lal R (2000) Physiological role of gap-junctional hemichannels: Extracellular calcium-dependent isosmotic volume regulation. Journal Of Cell Biology 148:1063–1074.

Quist AP, Chand A, Ramachandran S, Daraio C, Jin S, Lal R (2007) Atomic force microscopy imaging and electrical recording of lipid bilayers supported over microfabricated silicon chip nanopores: Lab-on-a-chip system for lipid membranes and ion channels. Langmuir 23:1375–1380.

Ramachandran S, Quist AP, Kumar S, Lal R (2006) Cisplatin nanoliposomes for cancer therapy: AFM and fluorescence Imaging of cisplatin encapsulation, stability, cellular uptake, and toxicity. 22:8156–8162.

Ramachandran S, Cohen D, Quist AP, Lal R (2008) New LED-based TIRF Microscopy To Study Molecular Permeability, Cell Membrane And Associated Cytoplasmic Dynamics. Biophys J 94:836.

Ramachandran S, Xie L-H, John SA, Subramaniam S, Lal R (2007) A Novel Role for Connexin Hemichannel in Oxidative Stress and Smoking-Induced Cell Injury. PLoS ONE 2:e712.

Sakmann B, Neher E (1983) Single Channel Recording. New York: Plenum.

Seo J, Ionescu-Zanetti C, Diamond J, Lal R, Lee LP (2004) Integrated multiple patch-clamp array chip via lateral cell trapping junctions. Applied Physics Letters 84:1973–1975.

Shroff SG, Saner DR, Lal R (1995) Dynamic Micromechanical Properties Of Cultured Rat Atrial Myocytes Measured By Atomic-Force Microscopy. American Journal Of Physiology-Cell Physiology 38:C286-C292.

Sulchek T, Yaralioglu GG, Quate CF, Minne SC (2002) Characterization and optimization of scan speed for tapping-mode atomic force microscopy. Review Of Scientific Instruments 73:2928–2936.

Thimm J, Mechler A, Lin H, Rhee S, Lal R (2005) Calcium-dependent open/closed conformations and interfacial energy maps of reconstituted hemichannels. Journal Of Biological Chemistry 280:10646–10654.

Thompson JB, Kindt JH, Drake B, Hansma HG, Morse DE, Hansma PK (2001) Bone indentation recovery time correlates with bond reforming time. Nature 414:773–776.

Wee KW, Kang GY, Park J, Kang JY, Yoon DS, Park JH, Kim TS (2005) Novel electrical detection of label-free disease marker proteins using piezoresistive self-sensing micro-cantilevers. Biosensors & Bioelectronics 20:1932–1938.

Weiner S, Wagner HD (1998) The material bone: Structure mechanical function relations. Annual Review Of Materials Science 28:271–298.

Worcester DL, Kim HS, Miller RG, Bryant PJ (1990) Imaging Bacteriorhodopsin Lattices In Purple Membranes With Atomic Force Microscopy. Journal Of Vacuum Science & Technology A-Vacuum Surfaces And Films 8:403–405.

Xu J, Wang XB, Ensign B, Li M, Wu L, Guia A, Xu JQ (2001) Ion-channel assay technologies: quo vadis? Drug Discovery Today 6:1278–1287.

Yuan CB, Chen A, Kolb P, Moy VT (2000) Energy landscape of streptavidin-biotin complexes measured by atomic force microscopy. Biochemistry 39:10219–10223.

Zhang XH, Wojcikiewicz E, Moy VT (2002) Force spectroscopy of the leukocyte function-associated antigen-1/intercellular adhesion molecule-1 interaction. Biophysical Journal 83:2270–2279.

# Part III
# Nanobiotechnology: Biologically Inspired Nanoengineering and Their Applications

# Molecular Motors and Machines

**Serena Silvi and Alberto Credi**

## 1 Introduction

Movement is one of life's central attributes. Nature provides living systems with complex molecules called motor proteins, which work inside a cell like ordinary machines built for everyday needs. The development of civilization has always been strictly related to the design and construction of devices, from wheel to jet engine, capable of facilitating man movement and traveling. Nowadays, the miniaturization race leads scientists to investigate the possibility of designing and constructing motors and machines at the nanometer scale, i.e., at the molecular level. Chemists, by the nature of their discipline, are able to manipulate atoms and molecules and are therefore in the ideal position to develop bottom-up strategies for the construction of nanoscale devices.

Natural molecular motors are extremely complex systems; their structures and detailed working mechanisms have been elucidated only in a few cases and any attempt to construct systems of such a complexity by using the bottom-up molecular approach would be hopeless. What can be done, at present, in the field of artificial molecular motors is to construct simple prototypes consisting of a few molecular components, capable of moving in a controllable way, and to investigate the challenging problems posed by interfacing artificial molecular devices with the macroscopic world, particularly as far as energy supply and information exchange are concerned. Surely, the study of motion at the molecular level is a fascinating topic from the viewpoint of basic research and a promising field for novel applications.

In the first section of this chapter, we shall introduce the concepts of molecular motors and machines, and we describe the bottom-up (i.e., supramolecular) approach to their construction. We then discuss the characteristics of molecular motors, with

S. Silvi • A. Credi (✉)
Dipartimento di Chimica "G. Ciamician", Università di Bologna,
Bologna, Italy
e-mail: alberto.credi@unibo.it

G.A. Silva and V. Parpura (eds.), *Nanotechnology for Biology and Medicine: At the Building Block Level*, Fundamental Biomedical Technologies,
DOI 10.1007/978-0-387-31296-5_4, © Springer Science+Business Media, LLC 2012

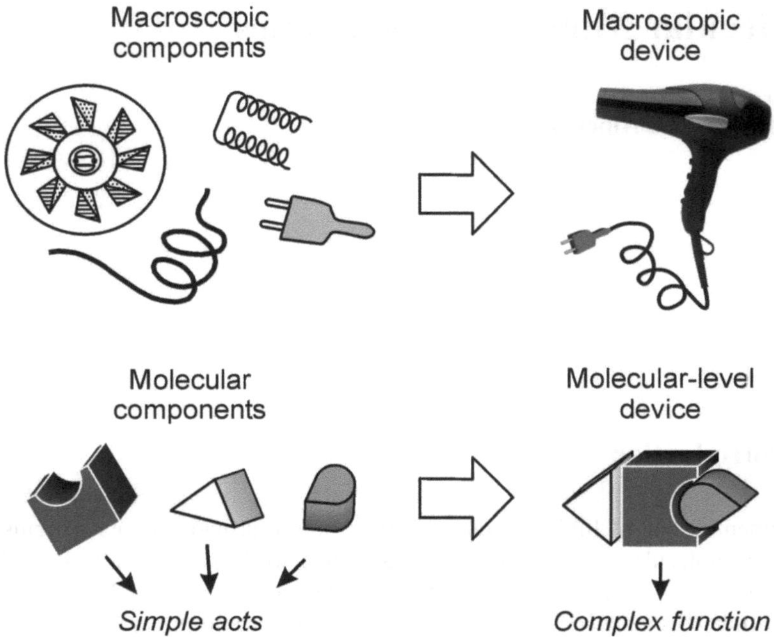


**Fig. 1** Extension of the concept of macroscopic device to the molecular level. Reproduced by permission from Balzani et al. 2003

reference to those of macroscopic ones. In the second part, we briefly illustrate the working principles of some natural molecular motors. The third section deals with artificial molecular motors and machines. For space reasons, we shall describe only selected examples, often taken by our own research. However, other relevant contributions in this field are mentioned in the text or listed in the references. A few hybrid systems, i.e., obtained either by suitable engineering of motor proteins or by integrating natural and artificial molecular devices, are illustrated in the fourth part. Finally, we critically discuss some perspectives and limitations of artificial molecular motors and machines.

## *1.1 The Bottom-Up (Supramolecular) Approach to Nanodevices*

In everyday life, we make extensive use of devices. A *device* is something invented and constructed for a special purpose. More specifically, it is an assembly of components designed to achieve a specific *function*, resulting from the cooperation of the (simple) *acts* performed by each component (Fig. 1, top). A *machine* is a particular type of device in which the component parts display changes in their relative positions as a result of some external stimulus.

Depending on the purpose of its use, a device can be very big or very small. In the last 50 years, progressive miniaturization of the components employed for the construction of devices and machines has resulted in outstanding technological achievements, particularly in the field of information processing. A common prediction is that further progress in miniaturization will not only decrease the size and increase the power of computers, but could also open the way to new technologies in the fields of medicine, environment, energy, and materials.

Until now, miniaturization has been pursued by a large-downward (top-down) approach, which is reaching practical and fundamental limits (Keyes 2001). Miniaturization, however, can be pushed further on since "there is plenty of room at the bottom," as Richard Feynman stated in a famous talk to the American Physical Society in 1959 (Feynman 1960a, b). The key sentence of Feynman's talk was the following: "The principles of physics do not speak against the possibility of manoeuvring things atom by atom." The idea of the "atom-by-atom," bottom-up approach to the construction of nanoscale devices and machines, however, which was so much appealing to some physicists (Drexler 1986, 1992), did not convince chemists who are well-aware of the high reactivity of most atomic species and of the subtle aspects of chemical bond. Chemists know (Smalley 2001) that atoms are not simple spheres that can be moved from a place to another place at will. Atoms do not stay isolated; they bond strongly to their neighbors and it is difficult to imagine that the atoms can be taken from a starting material and transferred to another material.

In the late 1970s, a new branch of chemistry, called *supramolecular chemistry*, emerged and expanded very rapidly. In the frame of research on supramolecular chemistry, the idea began to arise in a few laboratories (Joachim and Launay 1984; Lehn 1990; Balzani and Scandola 1991) that molecules are much more convenient building blocks than atoms to construct nanoscale devices and machines (Fig. 1, bottom). The main reasons at the basis of this idea are: (1) molecules are stable species, whereas atoms are difficult to handle; (2) nature starts from molecules, not from atoms, to construct the great number and variety of nanodevices and nanomachines that sustain life; (3) most of the laboratory chemical processes deal with molecules, not with atoms; (4) molecules are objects that exhibit distinct shapes and carry device-related properties (e.g., properties that can be manipulated by photochemical and electrochemical inputs); (5) molecules can self-assemble or can be connected to make larger structures. In the same period, research on molecular electronic devices began to flourish (Aviram and Ratner 1974; Carter 1982; Metzger 2003).

In the following years, supramolecular chemistry grew very rapidly (Lehn 1995, 1996; Steed and Atwood 2000, 2004) and it became clear that the bottom-up approach based on molecules opens virtually unlimited possibilities concerning design and construction of artificial molecular devices and machines. More recently, the concept of molecules as nanoscale objects exhibiting their own shape, size, and properties has been confirmed by new, very powerful techniques, such as single-molecule fluorescence spectroscopy and the various types of probe microscopies, capable of visualizing (Rigler et al. 2001; Moerner 2002; Zander et al. 2002) or manipulating (Gimzewski and Joachim 1999; Hlavina et al. 2001; Samori et al. 2004) single molecules, and even to investigate bimolecular chemical reactions at the single-molecule level (Christ et al. 2001).

Much of the inspiration to construct molecular devices and machines comes from the outstanding progress of molecular biology that has begun to reveal the secrets of the natural nanodevices which constitute the material base of life (Goodsell 2004). The bottom-up construction of devices as complex as those present in nature is, of course, an impossible task. Therefore, chemists have tried to construct much simpler systems without mimicking the complexity of the biological structures. In the last few years, synthetic talent that has always been the most distinctive feature of chemists, combined with a device-driven ingenuity evolved from chemists' attention to functions and reactivity, has led to outstanding achievements in this field (Balzani et al. 2000a, 2003; 2001a, b).

## *1.2 Characteristics of Molecular Motors and Machines*

A *molecular motor* can be defined as an assembly of a discrete number of molecular components designed to perform mechanical-like movements under the control of appropriate energy inputs (Akkerman and Coops 1967; Cozzi et al. 1981; Bedard and Moore 1995; Dominguez et al. 2003; Shima et al. 2004). The words *motor* and *machine* are often used interchangeably when referred to molecular systems. It should be recalled, however, that a motor converts energy into mechanical work while a machine is a device, usually containing a motor component, designed to accomplish a function. Molecular motors and machines operate via electronic and/or nuclear rearrangements and make use of thermal fluctuations (Brownian motion) (Astumian and Hänggi 2002; Parisi 2005). Like the macroscopic counterparts, they are characterized by (1) the kind of energy input supplied to make them work; (2) the type of motion (linear, rotatory, oscillatory, etc.) performed by their components; (3) the way in which their operation can be monitored; (4) the possibility to repeat the operation at will (cyclic process); and (5) the timescale needed to complete a cycle. According to the view described above, an additional and very important distinctive feature of a molecular machine with respect to a molecular motor is (6) the function performed (2001a).

### 1.2.1 Energy Supply

The problem of the energy supply to make artificial molecular motors work [point (1)] is of the greatest importance (Ballardini et al. 2001a, b). The most obvious way to supply energy to a chemical system is through an exergonic chemical reaction. In the previously mentioned address (Feynman 1960a, b) to the American Physical Society, Richard Feynman observed: "An internal combustion engine of molecular size is impossible. Other chemical reactions, liberating energy when cold, can be used instead." This is exactly what happens in our body, where the chemical energy supplied by food is used in long series of slightly exergonic reactions to power the biological machines that sustain life.

If an artificial molecular motor has to work by inputs of chemical energy, it will need addition of fresh reactants ("fuel") at any step of its working cycle, with the concomitant formation of waste products. Accumulation of waste products, however, compromises the operation of the device unless they are removed from the system, as it happens in our body as well as in macroscopic internal combustion engines. The need to remove waste products introduces noticeable limitations in the design and construction of artificial molecular motors based on chemical fuel inputs.

Chemists have since long known that photochemical and electrochemical energy inputs can cause the occurrence of *endergonic* and *reversible* reactions. In recent years, the outstanding progress made by supramolecular photochemistry (Balzani and Scandola 1991; Balzani 2003) and electrochemistry (Kaifer and Gómez-Kaifer 1999; Marcaccio et al. 2004) has, thus, led to the design and construction of molecular machines powered by light or electrical energy, which work without the formation of waste products. In the case of photoexcitation, the commonly used endergonic and reversible reactions are isomerization and redox processes. In the case of electrochemical energy inputs, the induced endergonic and reversible reactions are, of course, heterogeneous electron transfer processes. Photochemical and electrochemical techniques offer further advantages, since lasers provide the opportunity of working in very small space and very short time domains, and electrodes represent one of the best ways to interface molecular-level systems with the macroscopic world.

A very important feature of molecular motors, related to points (1) and (4), is their capability to exhibit an *autonomous* behavior; that is, to keep operating, in a constant environment, as long as the energy source is available. Natural motors are autonomous, but the vast majority of the artificial molecular motors reported so far are *not autonomous* since, after the mechanical movement induced by a given input, they need another opposite input to reset.

Needless to say, the operation of a molecular machine is accompanied by partial conversion of free energy into heat, regardless of the chemical, photochemical, and electrochemical nature of the energy input.

### 1.2.2 Other Features

The motions performed by the component parts of a molecular motor [point (2)] may imply rotations around covalent bonds or the making and breaking of intercomponent noncovalent bonds, as we shall see later on.

In order to control and monitor the device operation [point (3)], the electronic and/or nuclear rearrangements of the component parts should cause readable changes in some chemical or physical property of the system. In this regard, photochemical and electrochemical techniques are very useful since both photons and electrons can play the dual role of *writing* (i.e., causing a change in the system) and *reading* (i.e., reporting the state of the system). Luminescence spectroscopy, in particular, is a valuable technique since it is easily accessible and offers good sensitivity and selectivity, along with the possibility of time-resolved studies.

The operation timescale of molecular machines [point (5)] can range from microseconds to seconds, depending on the type of rearrangement and the nature of the components involved.

Finally, as far as point (6) is concerned, the functions that can be performed by exploiting the movements of the component parts in molecular machines are various and, to a large extent, still unpredictable. It is worth to note that the mechanical movements taking place in molecular-level machines, and the related changes in the spectroscopic and electrochemical properties, usually obey binary logic and can thus be taken as a basis for information processing at the molecular level. Artificial molecular machines capable of performing logic operations have been reported (Balzani 2003).

## 2 Natural Systems

In the last few years, much progress has been made in elucidation of the moving mechanisms of motor biomolecules, owing to the fact that – in addition to the established physiological and biochemical methods – novel in vitro techniques have been developed which combine optical and mechanical methods to observe the behavior of a single protein.

The most important and best-known natural molecular motors are adenosine triphosphate (ATP) synthases, myosin, and kinesin (2003; Schliwa and Woehlke 2003). An ATP synthase is the ubiquitous enzyme that manufactures ATP and is a rotary motor (Boyer 1993; Walker 1998). The enzymes of the myosin and the kinesin families are linear motors that move along polymer substrates (actin filaments for myosin and microtubules for kinesin), converting the energy of ATP hydrolysis into mechanical work (Frey 2002; Vale and Milligan 2000). Motion derives from a mechanochemical cycle, during which the motor protein binds to successive sites along the substrate in such a way as to move forward on average.

Several other biological processes are based on motions, including protein folding/unfolding. Another example is RNA polymerase, which moves along DNA while carrying out transcription, thus acting as a molecular motor. A short and simplified description of ATP synthase, myosin, and kinesin is reported below.

### *2.1 ATP Synthase*

This enzyme consists of two rotary molecular motors attached to a common shaft, each attempting to rotate in the opposite direction. The $F_1$ motor uses the free energy of ATP hydrolysis to rotate in one direction while the $F_0$ motor uses the energy stored in a transmembrane electrochemical gradient to turn in the opposite direction. Which motor "wins" (i.e., develops more torque) depends on cellular conditions. When $F_0$ takes over, which is the normal situation, it drives the $F_1$ motor in reverse whereupon it synthesizes ATP from its constituents, adenosine diphosphate (ADP)

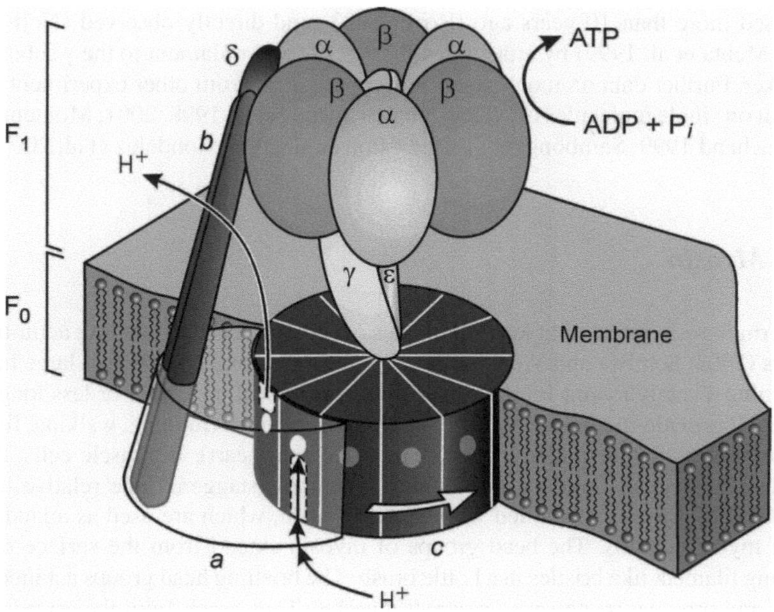


**Fig. 2** Schematic representation of the structure of $F_0F_1$–ATP synthase. See text for details. Reproduced by permission from Balzani et al. 2003

and inorganic phosphate, $P_i$. When $F_1$ dominates, it hydrolyzes ATP and drives the $F_0$ motor in reverse, turning it into an ion pump that moves ions across the membrane against the electrochemical gradient. The mechanochemistry of ATP synthase has been studied in great detail (Boyer 1993; Rastogi and Girvin 1999; Stock et al. 1999; Oster and Wang 2000; Ren and Allison 2000; Seelert et al. 2000; Bustamante et al. 2001; Frey 2002; 2003; Schliwa and Woehlke 2003) and new structural information continues to appear (Rondelez et al. 2005).

This enzyme consists of two principal domains (Fig. 2). The asymmetric membrane-spanning $F_0$ portion contains a proton channel, and the soluble $F_1$ portion contains three catalytic sites which cooperate in the synthetic reactions. The catalytic region is made up of nine protein subunits with the stoichiometry $3\alpha{:}3\beta{:}1\gamma{:}1\delta{:}1\varepsilon$, approximating to a flattened sphere, 10 nm across and 8 nm high. The flow of protons through $F_0$ generates a torque, which is transmitted to $F_1$ by an asymmetric shaft, the $\gamma$-subunit. This subunit acts as a rotating "cam" within $F_1$, sequentially releasing ATPs from the three active sites. The free energy difference across the inner membrane of mitochondria and bacteria is sufficient to produce 3 ATPs per 12 protons passing through the motor.

As mentioned above, the $F_0F_1$–ATP synthase is reversible, i.e., the full enzyme can synthesize or hydrolyze ATP; $F_1$ in isolation, however, can only hydrolyze it. The spinning of $F_1$–ATP synthase, i.e., the rotary motor nature of this enzyme, was first proposed more than 10 years ago (Boyer 1993) and directly observed (Noji et al. 1997; Mehta et al. 1999) by attaching a fluorescent actin filament to the $\gamma$ subunit as a marker. Further data on motor rotation were obtained from other experiments carried out on single molecules of ATP synthase (Yasuda et al. 1998, 2001; Montemagno and Bachand 1999; Sambongi et al. 1999; Itoh et al. 2004; Rondelez et al. 2005).

## 2.2 *Myosin*

The term myosin refers to at least 14 classes of proteins, each containing actin-based motors (2003; Schliwa and Woehlke 2003). Myosin is composed of two large heads, containing a catalytic unit for ATP hydrolysis, connected to a more or less long tail. Myosin-II provides the power for all our voluntary motions (running, walking, lifting, etc.) as well as for involuntary muscles (e.g., beating heart). In muscle cells, many myosin-II molecules combine by aligning their tails, staggered one relative to the next. Muscle cells are also filled with cables of actin, which are used as a ladder on which myosin climbs. The head groups of myosin extend from the surface of the resulting filament like bristles in a bottle brush. The bristling head groups act independently and provide the power to contract muscles. They reach from the myosin filament to a neighboring actin filament and attach to it. Breakage of an ATP molecule, then, forces the myosin head into a radically different shape. It bends near the center and drags the myosin filament along the actin filament. This is the power stroke of muscle contraction. In a rapidly contracting muscle, each myosin head may stroke five times a second, each stroke moving the filament about 10 nm. The working stroke of myosin-II was observed (Finer et al. 1994; Reconditi et al. 2004) by optical methods. These experiments, however, could not resolve a number of issues, such as the motor mechanism, which are still the subject of extensive investigations.

Unlike myosin-II, myosin-V is a processive motor, i.e., it moves progressively along its actin track, transporting organelles. In order to take consecutive steps along the actin filament without dissociating, the two heads must operate in a coordinated manner, but the mechanism has been controversial. Two models have been postulated: the “hand-over-hand” model, in which the two heads alternate in the lead (akin to human walking), and the “inchworm” model, in which the heads shuffle along one behind the other, with one of the two always leading. Recent detailed studies (Forkey et al. 2003; Molloy and Veigel 2003; Yildiz et al. 2003) strongly support a hand-over-hand mechanism of motility, and not an inchworm model.

## 2.3 *Kinesin*

Kinesin is a processive two-headed motor protein which moves along microtubule tracks. It transports a variety of cargo, including membranous organelles, mRNA, intermediate filaments, and signaling molecules. Models for the motility cycles of kinesin were elaborated (Endow and Higuchi 2000; Kikkawa et al. 2001; Svoboda et al. 1993; Visscher et al. 1999; 2003; Schliwa and Woehlke 2003) and the movement of single-kinesin molecules on microtubule tracks under variable loads and ATP concentrations was studied. A puzzling problem is the mechanism of directed motion along tracks that do not exhibit any chemical or electrical gradient. As for myosin-V, a hotly debated question (Hua et al. 2002) is whether the two heads of kinesin move in a hand-over-hand or in an inchworm manner. High-resolution fluorescence microscopy experiments (Yidiz et al. 2004) provided strong evidence in favor of the hand-over-hand model. Even if the intimate mechanism of the movement is not yet known, the fact remains that kinesin molecules generate force and transport cargo unidirectionally.

# 3 Artificial Systems

## *3.1 Rotary Motors*

Artificial molecular rotary motors are systems capable of undergoing unidirectional and repetitive rotations under the action of external energy inputs. The construction of molecular rotary motors poses several challenges particularly because it is difficult to satisfy the unidirectional rotation requirement. At present, artificial rotary motors have only been obtained by exploiting a C=C isomerization reaction in carefully designed hindered chiral alkenes (Sect. 3.1.1), but we also include the description of the chemically driven unidirectional rotation in helicene–triptycene compounds and of surface-mounted molecular rotors.

### 3.1.1 Rotary Motors Powered by Light Energy

Photochemical *trans–cis* isomerization around a carbon–carbon double bond is a well-know reversible reaction (Balzani and Scandola 1991). Exploitation of this photochemical reaction in a suitably designed compound which contains hindering substituents and stereogenic centers led to a photoinduced unidirectional rotary motion (Koumura et al. 1999).

In the alkene **1** shown in Fig. 3, each one of the two helical subunits linked by a double bond can adopt a right-handed (*P*) or a left-handed (*M*) helicity. As a result, a total of four stereoisomers are possible for this compound. The *cis–trans* photoisomerization reactions are reversible and occur upon irradiation at appropriate wavelengths. By contrast, the inversions of helicities while maintaining a *cis* or a *trans* configuration occur irreversibly under the influence of thermal energy. Upon irradiation (≥280 nm, 218 K) of a solution of (*P*,*P*)-*trans*-**1** (Fig. 3a), a mixture of (*P*,*P*)-*trans*-**1** and (*M*,*M*)-*cis*-**1** is obtained in a ratio of 5:95 (Fig. 3b). By warming the solution up to 293 K, (*M*,*M*)-*cis*-**1** interconverts irreversibly to (*P*,*P*)-*cis*-**1**

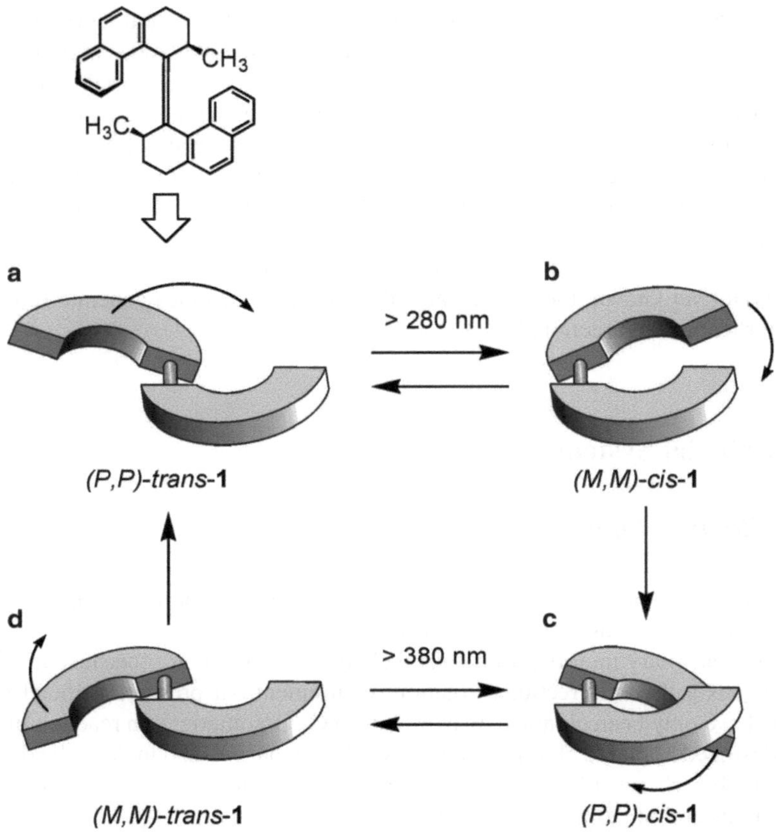


**Fig. 3** Light-driven unidirectional rotation of the two helical subunits of compound **1**. Reproduced by permission from Balzani et al. 2003

(Fig. 3c). Subsequent irradiation (≥280 nm) of the solution produces a mixture of (*P*,*P*)-*cis*-**1** and (*M*,*M*)-*trans*-**1** in a ratio of 10:90 (Fig. 3d). Upon increasing the temperature further (333 K), (*M*,*M*)-*trans*-**1** interconverts irreversibly to the original isomer (*P*,*P*)-*trans*-**1**. Thus, a sequence of light- and heat-induced isomerizations can be exploited to move this molecular rotor in one direction only. Indeed, when (*P*,*P*)-*trans*-**1** is irradiated (≥280 nm) at a high temperature (293 K), a clockwise 360° rotation occurs and therefore the motor exhibits autonomous behavior. The overall process can be followed by monitoring the changes in the circular dichroism spectra. The unidirectional motion in this system is dictated by the stereogenic centers associated with the two methyl substituents. More recently, this molecular motor was redesigned to improve its performance, (Ter Wiel et al. 2003) and the structure was modified so that the motor can be powered by visible instead of UV light (Van Delden et al. 2003). Furthermore, it was shown that unidirectional rotation can be controlled by a single stereogenic center (Koumura et al. 2000).

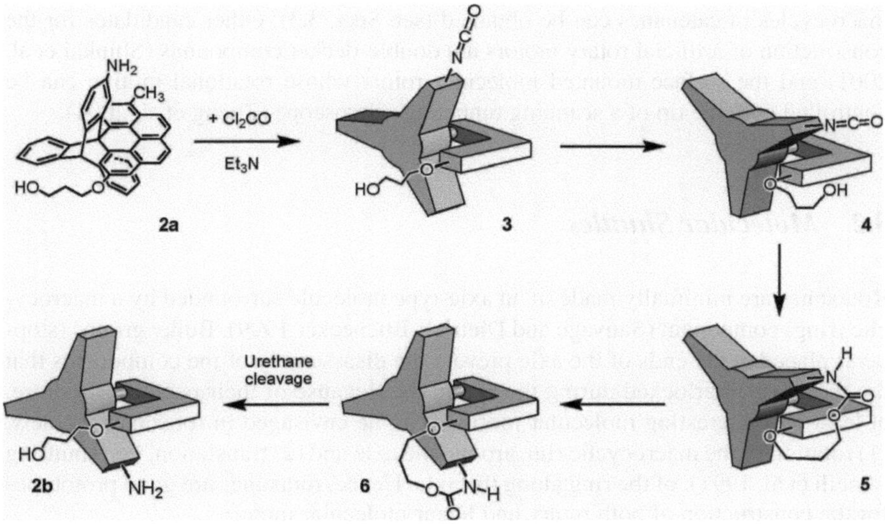


**Fig. 4** Sequence of events causing unidirectional rotation of 120° in a triptycene/helicene system powered by phosgene as a chemical fuel. Adapted by permission from Balzani et al. 2003

### 3.1.2 Toward a Chemically Driven Rotary Motor

An important step toward the realization of a chemically powered molecular motor is the unidirectional rotation illustrated in Fig. 4 (Kelly et al. 1999, 2000). Compounds **2a** and **2b** are two of the three low-energy rotamers about the axle connecting the triptycene and helicene components. Rotamer **2a** is activated by reaction with phosgene to give isocyanate **3**, which is chemically "armed" to react with the OH group in the hydroxypropyl tether attached to helicene. However, in the rotational ground state **3**, the isocyanate and the OH group are too far apart to interact. However, at those instants when a clockwise rotation of the triptycene (not possible with a comparable counterclockwise rotation) brings the isocyanate and the OH group sufficiently close to react (see **4**), urethane formation can then result (**5**), irreversibly trapping the triptycene in a relatively high energy conformation around the triptycene/helicene axle. Ambient thermal energy then drives the exergonic, but very slow, unidirectional rotation from **5** to **6**. Finally, **6** is cleaved to **2b**, thereby completing the chemically driven rotation of **2a** to **2b**. Admittedly, after this proof of principle, much work has still to be done to obtain a system that can undergo a full, continuous, and fast rotation.

### 3.1.3 Other Systems

It was shown (Raehm et al. 1999; Kern et al. 2000; Poleschak et al. 2004) that in suitably designed rotaxanes the pirouetting-type movements of the ring around the axle can be electrochemically driven. Similarly, controlled pirouetting of the interlocked macrocycles in catenanes can be obtained (see Sect. 3.3). Other candidates for the construction of artificial rotary motors are double-decker compounds (Shinkai et al. 2001) and the surface-mounted molecular rotors whose rotational motion can be controlled with the tip of a scanning tunneling microscope (Zheng et al. 2004).

## 3.2 *Molecular Shuttles*

Rotaxanes are minimally made of an axle-type molecule surrounded by a macrocyclic (ring) component (Sauvage and Dietrich-Buchecker 1999). Bulky groups (stoppers) placed at the ends of the axle prevent the disassembly of the components that are therefore interlocked during the synthesis. Because of their peculiar structure, at least two interesting molecular motions can be envisaged in rotaxanes, namely, (1) rotation of the macrocyclic ring around the axle and (2) translation, i.e., shuttling (Anelli et al. 1991), of the ring along the axle. Hence, rotaxanes are good prototypes for the construction of both rotary and linear molecular motors.

The axle and ring components usually exhibit some kind of interaction originating from complementary chemical properties, which is also exploited in the template-directed synthesis of such systems. In rotaxanes containing two different recognition sites in the axle component, it is possible to switch the position of the ring between the two "stations" by an external stimulus. Systems of this type, termed molecular shuttles, constitute indeed the most common implementation of the molecular motor concept with rotaxanes.

### 3.2.1 Molecular Shuttles Powered by Chemical Reactions

The first example of a chemically driven molecular shuttle was reported in 1994 (Bissell et al. 1994). Since then, many molecular shuttles relying on chemical stimulation have been described in the literature (Balzani et al. 2000a; 2003; 2001a, b; Elizarov et al. 2002; Tseng et al. 2003; Keaveney and Leigh 2004; Leigh and Perez 2004). One of the best systems in terms of switching and stability is compound $\mathbf{7}^{3+}$ shown in Fig. 5 (Ashton et al. 1998). It is made of an axle component containing an ammonium (AMH) and an electron acceptor bipyridinium (BPM) unit that can establish hydrogen-bonding and electron donor-acceptor interactions, respectively, with the ring component, which is a crown ether with electron-donor properties. An anthracene moiety is used as a stopper because its absorption, luminescence, and redox properties are useful to monitor the state of the system. Since the $N^+$–H···O hydrogen bonding interactions between the macrocycle and the ammonium center are much stronger than the electron donor-acceptor interactions of the macrocycle with the BPM, the rotaxane exists as only one of the two possible translational isomers (Fig. 5a, state 0). Deprotonation of the ammonium center with a base (Fig. 5b) causes 100% displacement of the macrocycle by Brownian motion to the BPM unit (Fig. 5c, state 1); reprotonation with an acid (Fig. 5d) directs the macrocycle back

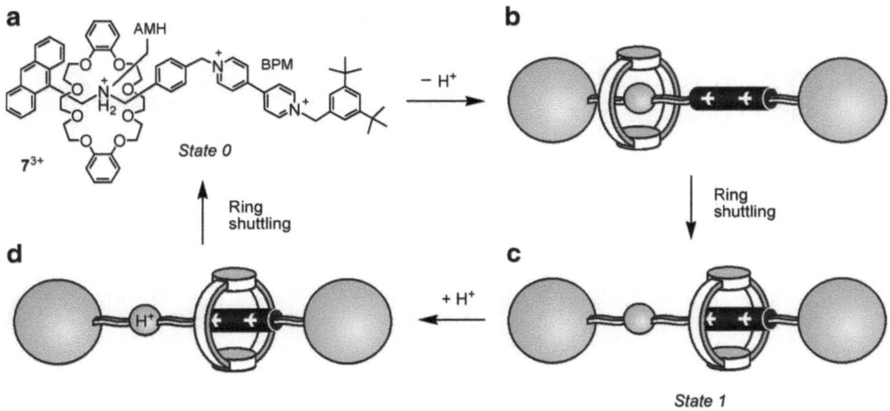


**Fig. 5** Schematic representation of the operation of the acid/base-controllable molecular shuttle $\mathbf{7}^{3+}$. Adapted by permission from Balzani et al. 2003

onto the ammonium center. Such a switching process was investigated in solution by nuclear magnetic resonance (NMR) spectroscopy and by electrochemical and photophysical measurements (Ashton et al. 1998). Recently, the kinetics of ring shuttling were also studied in detail by stopped-flow spectroscopic experiments (Garaudée et al. 2008). The full chemical reversibility of the energy supplying acid/base reactions guarantees the reversibility of the mechanical movement, in spite of the formation of waste products. Notice that this system could be useful for information processing since it is a bistable system and can exhibit a binary logic behavior. It should also be noted that, in the deprotonated rotaxane, it is possible to displace the ring from the bipyridinium station by destroying the donor–acceptor interaction through reduction of the bipyridinium station. Therefore, in this system, mechanical movements can be induced by two different types of stimuli (acid base and electron hole).

### 3.2.2 Molecular Shuttles Powered by Electrical Signals

A good example of an electrochemically driven molecular shuttle is rotaxane **8** reported in Fig. 6. It consists of a benzylic amide macrocycle that surrounds an axle featuring two hydrogen-bonding stations, namely, a succinamide (SA) and a naphthalimide (NI) unit, separated by a long alkyl chain (Brouwer et al. 2001; Altieri et al. 2003). Initially, the ring resides onto the SA station (Fig. 6a) because the NI unit is a much poorer hydrogen-bonding recognition site. Electrochemical reduction of the NI unit to the radical anion species can be carried out at −1.40 V versus the saturated calomel electrode (Fig. 6b). Since the naphthalimide anion is a much stronger hydrogen-bonding station compared to the succinamide, thermal fluctuations drive the ring from the latter to the former station (Fig. 6c). Successive electrochemical oxidation (−0.90 V) of the naphthalimide anion back to the neutral state (Fig. 6d) is

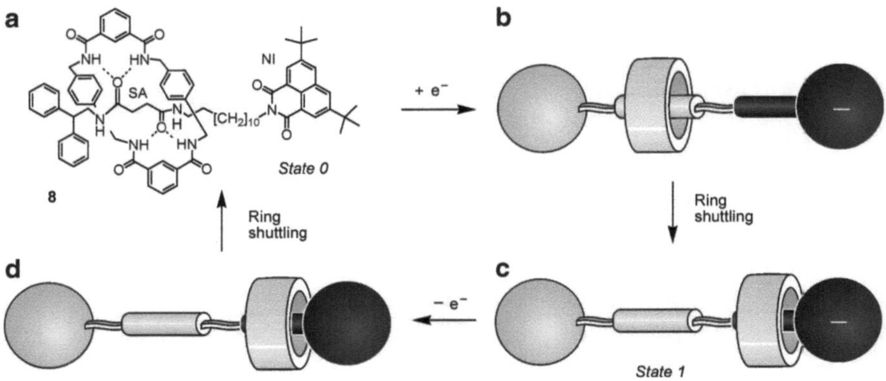


**Fig. 6** Schematic representation of the operation of the electrochemically driven molecular shuttle 8

followed by ring shuttling to the original SA station. The occurrence of these processes was demonstrated by cyclic voltammetric experiments and confirmed by flash photolysis. Additionally, these techniques provided information on the dynamics of ring shuttling. Other kinds of electrochemically driven molecular shuttles have been reported (Balzani et al. 2000a; 2003; 2001a, b; Tseng et al. 2004).

### 3.2.3 Molecular Shuttles Powered by Light Energy

The design and construction of molecular shuttles powered exclusively by light energy is a fascinating yet challenging subject. Rotaxane $\mathbf{9}^{6+}$ (Fig. 7) was specifically designed to achieve photoinduced ring shuttling. This compound is made of the electron donor macrocycle R, an axle component which contains a ruthenium(II) polypyridine complex (P) as one of its stoppers, a *p*-terphenyl-type rigid spacer (S), a 4,4′-bipyridinium unit ($A_1$), and a 3,3′-dimethyl-4,4′-bipyridinium unit ($A_2$) as electron accepting stations, and a tetraarylmethane group as the second stopper (T). The stable translational isomer of rotaxane $\mathbf{9}^{6+}$ (state 0) is the one in which the R component encircles the $A_1$ unit, in keeping with the fact that this station is a better electron acceptor than the other one. The strategy devised in order to obtain the photoinduced abacus-like movement of the R macrocycle between the two stations $A_1$ and $A_2$, illustrated in the bottom part of Fig. 7, is based on the following four operations.

(a) *Destabilization of the stable translational isomer*: Light excitation of the photoactive unit P (process 1) is followed by the transfer of an electron from the excited state to the $A_1$ station, which is encircled by the ring R (process 2), with the consequent "deactivation" of this station; such a photoinduced electron transfer process has to compete with the intrinsic decay of *P (process 3).
(b) *Ring displacement*: The ring moves by Brownian motion (process 4) from the reduced $A_1$ station to $A_2$ (state 1), a step that has to compete with the back electron

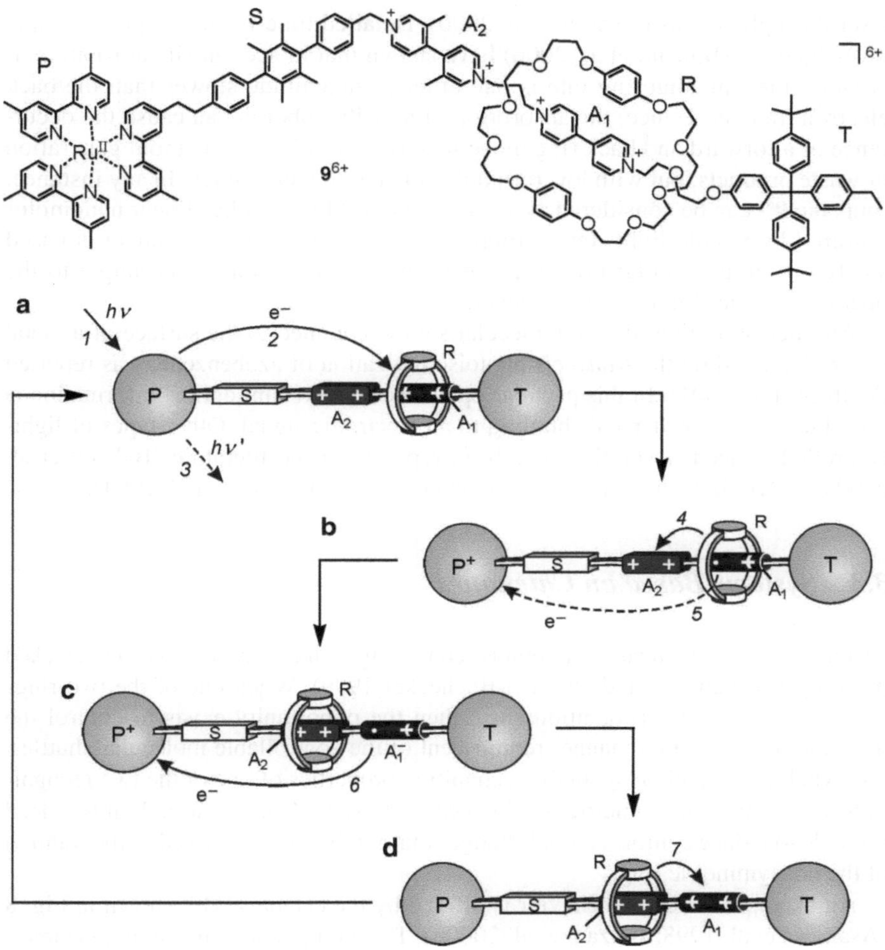


**Fig. 7** Structural formula of the rotaxane $\mathbf{9}^{6+}$ (*top*) and schematic representation of its operation as an autonomous "four-stroke" linear nanomotor powered by light (*bottom*)

transfer process from $A_1^-$ (still encircled by R) to the oxidized photoactive unit, $P^+$ (process 5).

(c) *Electronic reset*: A back electron transfer process from the "free" $A_1^-$ station to $P^+$ (process 6) restores the electron acceptor power to the $A_1$ station.

(d) *Nuclear reset*: As a consequence of the electronic reset, back movement of the ring by Brownian motion from $A_2$ to $A_1$ takes place (process 7).

The crucial point for such a mechanism is indeed the favorable competition between ring displacement (process 4) and back electron transfer (process 5). The rate constants of the relevant electron transfer processes were measured by laser flash photolysis (Ashton et al. 2000). Detailed time-resolved spectroscopic investigations (Balzani et al. 2006) have shown that in acetonitrile at room temperature the ring shuttling rate is one order of magnitude slower than the back electron transfer. Hence, the absorption of a visible photon can cause the occurrence of a forward and back ring movement (i.e., a full cycle) without generation of waste products, but with low (around 2%) quantum efficiency. In any instance, rotaxane $\mathbf{9}^{6+}$ can be considered as an autonomous "four-stroke" linear nanomotor powered by visible light. Interestingly, since the shuttling mechanism is based solely on intramolecular processes, there are no limitations of principle to the operation at the single-molecule level.

An interesting light-driven molecular shuttle confined to the surface of a metal electrode, based on the *trans–cis* photoisomerization of azobenzene, was reported (Willner et al. 2001). In this peculiar optoelectronic system, optical information is transduced by a mechanical shuttling to an electronic signal. Other types of light-controlled molecular shuttles have been reported in the literature (Balzani et al. 2000a; 2001b; Balzani et al. 2003; Abraham et al. 2004; Wang et al. 2004).

## *3.3 Systems Based on Catenanes*

Catenanes are chemical compounds consisting minimally of two interlocked macrocycles (Sauvage and Dietrich-Buchecker 1999). When one of the two rings carries two different recognition sites, then the opportunity exists to control the dynamic processes in a manner reminiscent of the controllable molecular shuttles. By switching *off* and *on* again the recognition properties of one of the two recognition sites of the nonsymmetric ring by means of external energy stimuli, it is indeed possible to induce conformational changes that can be simply viewed as the rotation of the nonsymmetric ring.

An example of such a behavior is offered by the catenane $\mathbf{10}^{4+}$ shown in Fig. 8 (Asakawa et al. 1998; Balzani et al. 2000b). This compound is made of a symmetric tetracationic ring containing two electron acceptor BPMs and a nonsymmetric ring comprising two different electron donor units, namely, a tetrathiafulvalene (TTF) group and a 1,5-dioxynaphthalene (DMN) unit. Since the TTF unit is by far a better electron donor than the DMN one, the thermodynamically stable conformation of the catenane is that in which the symmetric ring encircles the TTF unit of the nonsymmetric one (Fig. 8a, state 0). On electrochemical oxidation in solution, the TTF unit loses its electron donor power and acquires a positive charge (Fig. 8b). As a consequence, it is expelled from the cavity of the tetracationic ring and is replaced by the neutral DMN unit (Fig. 8c, state 1). At this stage, subsequent reduction restores the TTF unit (Fig. 8d) and the system goes back to its original conformation. A variety of techniques, including cyclic voltammetry, were employed to characterize the system. This compound was incorporated in a solid-state device that can be potentially used for random access memory (RAM) storage (Collier et al. 2000; Luo et al. 2002).

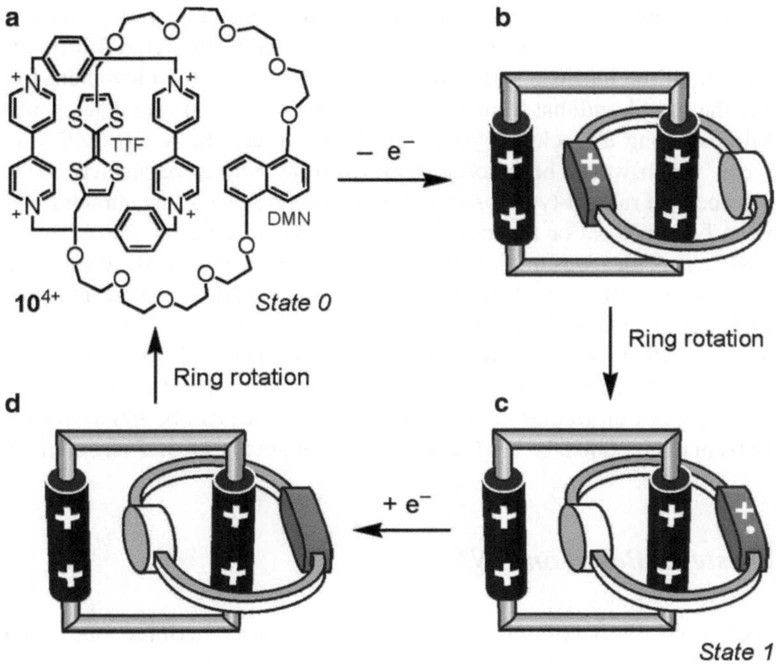


**Fig. 8** Redox-controlled ring rotation in catenane $\mathbf{10^{4+}}$

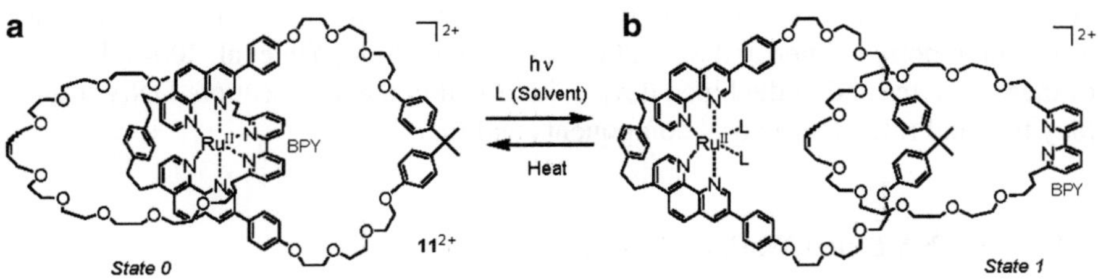


**Fig. 9** Photochemically and thermally induced motions taking place in catenane $\mathbf{11^{2+}}$

A clever strategy to obtain the light-driven rotation of the molecular rings in a catenane was reported (Mobian et al. 2004). Catenane $\mathbf{11^{2+}}$ (Fig. 9a, State 0) is essentially a ruthenium(II) complex designed in order that visible excitation in acetonitrile solution leads to the formation of metal-centered dissociative excited states which, in turn, cause the decoordination of the bipyridine (BPY) ligand. As a result, the rings mutually rotate to give the structure shown in Fig. 9b (state 1). Simple heating regenerates the starting complex, with both reactions (decoordination/recoordination) being quantitative. The overall process was monitored by NMR and UV–vis spectroscopy.

It should be pointed out that in the catenane systems described above repeated switching between the two states does not need to occur through a full rotation. In fact, because of the intrinsic symmetry of the system, both the movement from the state 0 to the state 1 and that from the state 1 to the state 0 can take place, with equal probabilities, along a clockwise or anticlockwise direction. A full (360°) rotation movement, which would be much more interesting from a mechanical viewpoint, can only occur in ratchet-type systems, i.e., in the presence of asymmetry elements which can be structural or functional in nature (Ballardini et al. 2001a, b; Balzani et al. 2003). This idea was implemented with a carefully designed catenane by relying on a sequence of photochemical, chemical, and thermally activated processes (Hernández et al. 2004). NMR spectroscopy was employed to characterize the system. A clever, albeit complex, way to obtain a unidirectional full rotation in a catenane made of three interlocked macrocycles was also devised (Leigh et al. 2003). Other examples of molecular motors based on catenanes can be found in the literature (Balzani et al. 2000a, b, 2003; 2001b; Abraham et al. 2004; Wang et al. 2004).

## 3.4 Systems Based on DNA

The DNA molecule is not only the repository of genetic heritage, but also a very interesting engineering material for nanotechnology (Niemeyer 1997, 2000). Moreover, it can be used for computation (Adleman 1994). Tweezers and other types of nanomachines based on conformational changes in DNA molecules have been described in the past few years (Mao et al. 1999; Yurke et al. 2000; Li and Tan 2002; Yan et al. 2002). DNA-based motors exhibiting autonomous behavior have also been reported (Chen and Mao 2004; Chen et al. 2004; Yin et al. 2004). In some of these systems and in the biped device described in the next section, DNA strands play the dual role of structural components and fuel.

### 3.4.1 A DNA Biped Walking Device

This complex device (Fig. 10) consists of two components: a "track" comprising three "stations" and two leg components, connected by flexible linkers, walking on the track (Sherman and Seeman 2004). Each station of the track and each leg of the biped terminate with single-stranded DNA portions, called footholds (A, B, and C) and feet (1 and 2), respectively, that are available to pair with complementary strands of DNA. The sequences of the feet and footholds have been carefully selected to minimize complementarities between them. A foot attaches to a foothold when a set strand S complementary *to both* is added to the solution. Each of these linking strands has an 8-base overhand or "toehold" which is not complementary to any of the feet or footholds. The toehold allows the set strand to be removed by pairing with a successively added unset strand U and the system is designed so that after the unset procedure all the unset strands and the set strands with which they are paired

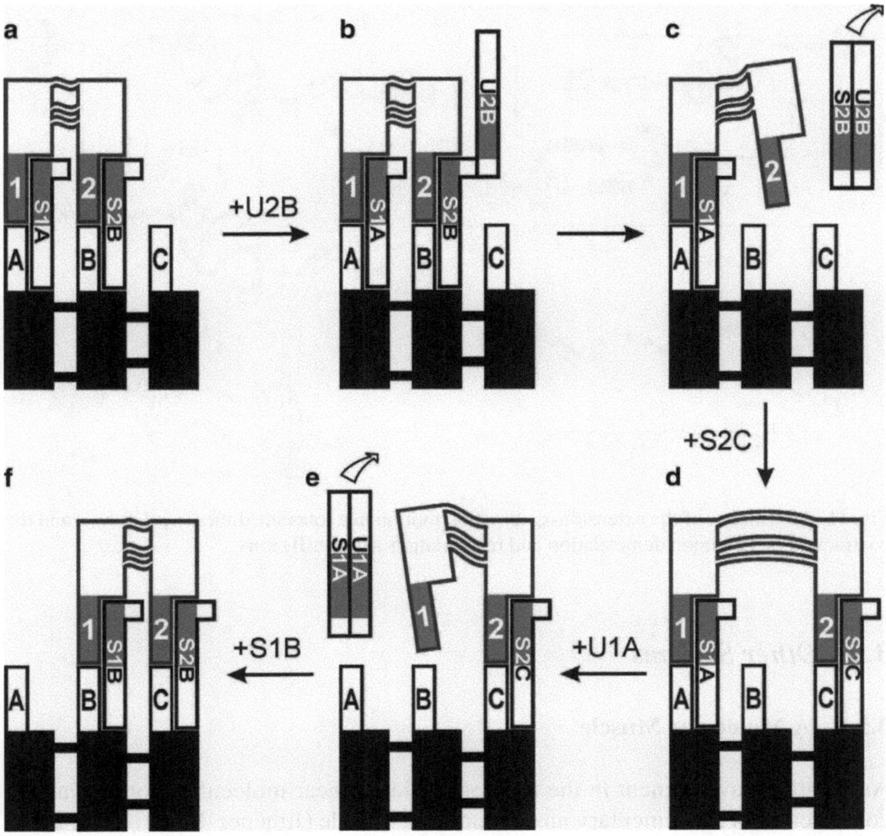


**Fig. 10** Cartoon representation of a full step of a DNA biped walking device

can be removed from the solution. The operation of the biped is illustrated in Fig. 10. At the beginning (a), the strand set S1A links the foothold A with the foot 1, and the set strand S2B links the foothold B with the foot 2. The first step consists of releasing the foot 2 from the foothold B, which is obtained by introducing in the solution the unset strand U2B which binds to the toehold of S2B (b) followed by branch migration that results in the complete hybridization of S2B with U2B (c). Once the foot 2 has been freed from the foothold B, it is free to be set to the foothold C by the appropriate set strand S2C (d). At this stage, the foot 1 can be freed from the foothold A by the appropriate unset strand U1A (e) and set to the foothold B by the appropriate set strand S1B (f). The state of the system is observed by taking an aliquot of the solution and exposing it to UV light that causes a crosslink reaction. Subsequent analysis by denaturation allows to establish which feet were attached to which footholds. Because of the relatively short lengths of the flexible linkers that connect the two legs, the biped moves by an "inchworm" mechanism.

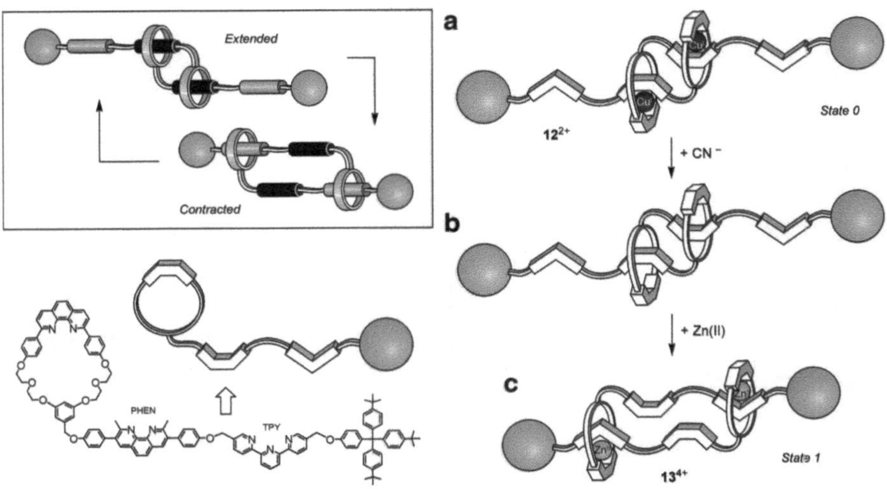


**Fig. 11** Schematics of the extension–contraction motion in a rotaxane dimer (*top left box*) and the contraction of $\mathbf{12}^{2+}$ upon demetalation and remetalation with Zn(II) ions

## 3.5 *Other Systems*

### 3.5.1 A Molecular Muscle

An exciting development in the field of artificial linear molecular motors was the construction of a rudimentary molecular-scale muscle (Jiménez-Molero et al. 2000, 2002). The idea started from the topology of a rotaxane dimer which, if suitably designed, can undergo contraction and stretching movements, as schematized in the box of Fig. 11. The synthesized system (Fig. 11) is a rotaxane dimer, $\mathbf{12}^{2+}$, that contains two copper(I) metal ions and two identical ring-and-axle components. Each of these components consists of a macrocyclic ring containing a bidentate phenanthroline unit, an axle containing a bidentate phenanthroline (PHEN), a terdentate terpyridine (TPY), and a bulky stopper unit. Each one of the two Cu(I) metal ions present in the rotaxane dimer is coordinated to two bidentate chelates since Cu(I) prefers a four-coordination arrangement. Under these conditions, the system is "extended" (Fig. 11a, state 0; length 8.3 nm, estimated from molecular models). Upon electrochemical oxidation of Cu(I), it was expected that the system contracts since Cu(II) prefers a five-coordination arrangement and therefore should be surrounded by a bidentate and a terdentate ligand. This change in the coordination environment, in fact, had been previously observed for rotaxanes and catenanes of the same family (Dietrich-Buchecker et al. 2003). In the case of $\mathbf{12}^{2+}$, however, electrochemical oxidation had no effect and contraction could be obtained only by chemical extraction of Cu(I) with a large excess of cyanide ions (Fig. 11b) and successive remetalation of the free ligand with zinc(II) ions (Fig. 11c, state 1).

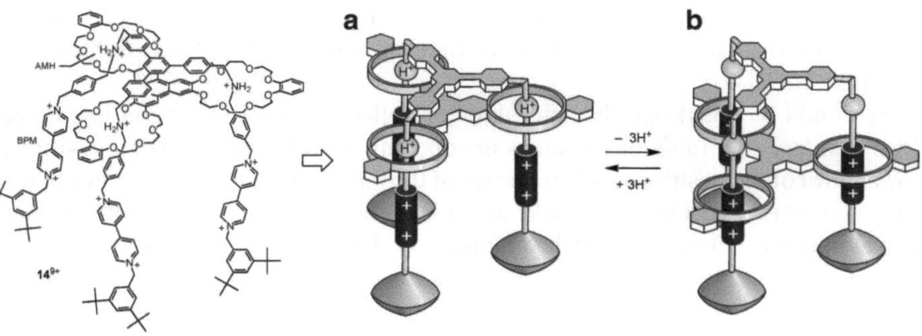


**Fig. 12** Chemical formula and operation scheme of a molecular elevator

The transformation of the contracted $\mathbf{13}^{4+}$ species – whose length was estimated to be 6.5 nm – to the extended $\mathbf{12}^{2+}$ one could be obtained by addition of Cu(I) ions in acetonitrile/dichloromethane solution.

### 3.5.2 A Molecular Elevator

The structural and functional features of the acid base-controllable molecular shuttle described in Sect. 3.2.1 were employed to design and construct a two-component molecular device, $\mathbf{14}^{9+}$ (Fig. 12), that behaves like a nanometer-scale elevator (Badjic et al. 2004). This nanomachine, which is ca. 2.5 nm in height and has a diameter of ca. 3.5 nm, consists of a trifurcated rig-like component containing two different notches – one AMH and one BPM – at different levels in each of its three legs. The latter are interlocked by a molecule made up of three crown ether macrocycles fused trigonally to a central aromatic floor; such a platform can be made to stop at the two different levels. The three legs of the rig carry bulky feet that prevent the loss of the platform. Initially, the platform resides exclusively on the "upper" level, i.e., with the three macrocycles surrounding the ammonium centers (Fig. 12a, state 0). This preference results from strong $N^+$–H···O hydrogen bonding and weak stabilizing $\pi$–$\pi$ stacking forces between the aromatic cores of the platform and rig components. Upon addition of a strong, non-nucleophilic phosphazene base to an acetonitrile solution of $\mathbf{14}^{9+}$, deprotonation of the ammonium center occurs and, as a result, the platform moves to the "lower" level, that is, with the three macrocycles surrounding the BPM (Fig. 12b, state 1). This structure is stabilized by electron donor–acceptor interactions between the electron-rich aromatic units of the platform and the electron-deficient BPM of the rig component. Subsequent addition of acid restores the AMH, and the platform moves back to the upper level. The elevator motion, which can be followed by NMR spectroscopy, electrochemistry, and absorption and fluorescence spectroscopy, is quantitative and can be repeated many times on the same solution. The distance traveled by the platform is about 0.7 nm, and from thermodynamic considerations it can be estimated that the elevator movement from the upper to lower level could in principle generate a force of up to 200 pN, one order of magnitude higher than that developed by myosin and kinesin (2003; Schliwa and Woehlke 2003).

It should be noted that the acid base-controlled mechanical motion in **14**$^{9+}$ can lead to interesting functions, such as the opening and closing of a large cavity and the control of the positions and properties of the bipyridinium legs. Since this behavior can be used to control the uptake and release of a guest molecule from the nanoelevator, it is of interest for the development of drug delivery systems.

## 4 Hybrid Systems

Recent scientific advances in molecular biology, supramolecular chemistry, and nanofabrication techniques have opened up the possibility of building functional hybrid devices based on natural motors. One long-term objective of this research is to utilize the finest attributes associated with the worlds of both biological and synthetic materials for the creation of nanomechanical systems that are powered by biological motors (Xi et al. 2005).

Hybrid molecular shuttle systems based on kinesin and microtubules (Sect. 2.3) were constructed (Hess et al. 2004) by using two different approaches: either the microtubules are fixed to a surface and kinesin is moving similar to cars moving on a highway and transporting kinesin-coated objects (Limberis and Stewart 2000) or the kinesin is bound to the surface and the microtubules are propelled by the kinesin analogous to a linear motor (Hess et al. 2001). An interesting application of the second technique is a statistical approach to surface imaging using fluorescent microtubules moving as probe robots across a surface coated with kinesin (Hess et al. 2002).

### *4.1 Light-Driven Production of ATP Within an Artificial Photosynthetic Membrane*

A multicomponent molecular system (tetrad) specifically designed to achieve photoinduced charge separation was vectorially incorporated in the bilayer membrane of a liposome. It was demonstrated (Steinberg-Yfrach et al. 1997) that, by visible light irradiation of this system under proper experimental conditions, protons were pumped into the liposome from the outside water solution. In principle, such a proton-motive force generated by photons can be utilized to perform mechanical work. This result was beautifully achieved (Steinberg-Yfrach et al. 1998; Gust et al. 2001) by incorporating $F_0F_1$–ATP synthase, with the ATP-synthesizing portion extending out into the external aqueous solution in liposomes containing the components of the proton-pumping photocycle. Irradiation of the membrane with visible light leads to the charge separation process that causes proton translocation with generation of a proton-motive force. On accumulation of sufficient proton-motive force, protons flow through the $F_0F_1$–ATP synthase, with the formation of ATP from ADP and $P_i$ (see Sect. 2.1).

The above-described system is the first complete biomimetic one which effectively couples electrical potential derived from photoinduced electron transfer to the chemical potential associated with the ADP–ATP conversion, thereby mimicking the entire process of bacterial photosynthesis. It constitutes a synthetic biological motor that, in principle, can be used to power anything which requires a proton gradient or ATP to work, or even future nanomachines. It might also be advantageous to use artificial, light-driven systems to produce ATP in order to carry out enzymatic reactions in the absence of interfering biological materials and without the need of living cells (Gust et al. 2001).

### *4.2 A Hybrid Nanomechanical Device Powered by ATP Synthase*

An interesting nanomechanical device was obtained by coupling nanofabrication techniques with biochemical engineering of a motor protein (Soong et al. 2000). Such nanomechanical device consists of three elements, namely: nanofabricated substrates of nickel posts, each 50–120 nm in diameter and 200-nm high; $F_1$–ATP synthase molecules, specifically modified for selective interface with the nanofabricated structures, nanofabricated Ni rods (150 nm in diameter and 750–1,500-nm long). The $F_1$–ATP synthase molecules were attached to the $Ni^{2+}$ of partially oxidized Ni posts using histidine tags introduced into the β-subunits (Bachand et al. 2001). Streptavidin was bound to the biotin residue on the γ-subunit tip, and the Ni nanorods, coated with biotinylated histidine-rich peptides, were then attached to the substrate-mounted $F_1$–ATP synthase motors through a biotin–streptavidin linkage. Rotation of the nanopropellers, which was observed in a flow cell with a CCD camera, was initiated by addition of ATP and inactivated by addition of $NaN_3$, an $F_1$–ATP synthase inhibitor. Although only 5 of 400 total observed propellers were found to rotate probably because of incorrect assembly of the components, these experiments demonstrate the possibility to integrate biomolecular motors with nanoengineered systems to produce nano- or micromechanical machines. From a chemical point of view, however, it should be noted that this device, unlike that described in Sect. 4.1, does not contain any artificial active component.

## 5 Conclusion and Perspectives

The results described here show that, by taking advantage of careful incremental design strategies of the tools of modern synthetic chemistry, of the paradigms of supramolecular chemistry, as well as of inspiration by natural systems, it is possible to produce compounds capable of performing nontrivial mechanical movements and exercising a variety of different functions upon external stimulation.

In the previously mentioned address to the American Physical Society (Feynman 1960a, b), R. P. Feynman concluded his reflection on the idea of constructing molecular machines as follows: "What would be the utility of such machines? Who knows? I cannot see exactly what would happen, but I can hardly doubt that when we have some control of the rearrangement of things on a molecular scale we will get an enormously greater range of possible properties that substances can have, and of different things we can do." This sentence, pronounced in 1959, is still an appropriate comment to the work described in this chapter. The results achieved enable to devise future developments, which are under investigation in our laboratory: (1) the design and construction of more sophisticated artificial molecular motors and machines; (2) the use of such systems to do tasks, such as molecular-level transportation, catalysis, and mechanical gating of molecular channels; and (3) the possibility of exploiting their logic behavior for information processing at the molecular level and, in the long run, for the construction of chemical computers.

It should also be noted that the majority of the artificial molecular motors developed so far operate in solution, that is, in an incoherent fashion and without control of spatial positioning. The solution studies of complicated chemical systems, such as molecular motors and machines, are indeed of fundamental importance to understand their operation mechanisms; moreover, for some use (e.g., drug delivery), molecular machines have to work in liquid solution. In this regard, it should be recalled that motor proteins operate in, or at least in contact with, an aqueous solution. However, it seems reasonable that, before artificial molecular motors and machines can find applications in many fields of technology, they have to be interfaced with the macroscopic world by ordering them in some way. The next generation of molecular machines and motors need to be organized at interfaces, deposited on surfaces, or immobilized into membranes or porous materials (Álvaro et al. 2003; Cavallini et al. 2003; Long et al. 2003; Cecchet et al. 2004; Flood et al. 2004; Hernandez et al. 2004; Huang et al. 2004; Katz et al. 2004) so that they can behave coherently and can be addressed in space. Indeed, the preparation of modified electrodes represents one of the most promising ways to achieve this goal.

Apart from more or less futuristic applications, the extension of the concept of motor and machine to the molecular level is of interest not only for the development of nanotechnology, but also for the growth of basic research. Looking at molecular and supramolecular species from the viewpoint of functions with references to devices of the macroscopic world is indeed a very interesting exercise, which introduces novel concepts into chemistry as a scientific discipline.

**Acknowledgments** We would like to thank professors Vincenzo Balzani and Margherita Venturi for stimulating discussions. Financial support from EU (STREP "Biomach" NMP2-CT-2003-505487), Ministero dell'Istruzione, dell'Università e della Ricerca (PRIN "Supramolecular Devices" and FIRB RBNE019H9K), and Università di Bologna (Funds for Selected Research Topics) is gratefully acknowledged.

# References


(2001a) Molecular Machines Special Issue. Acc Chem Res 34:409.

(2001b) Special Volume on Molecular Machines and Motors. Struct Bond 99:1.

(2003) Molecular Motors. Weinheim: Wiley-VCH.

Abraham W, Grubert L, Grummt U, Buck K (2004) A Photoswitchable Rotaxane with a Folded Molecular Thread. Chem Eur J 10:3562–3568.

Adleman L (1994) Molecular computation of solutions to combinatorial problems. Science 266: 1021–1024.

Akkerman O, Coops J (1967) Rec Trav Chim Pays-Bas 86:755.

Altieri A, Gatti F, Kay E, Leigh D, Martel D, Paolucci F, Slawin A, Wong J (2003) Electrochemically Switchable Hydrogen-Bonded Molecular Shuttles. J Am Chem Soc 125:8644–8654.

Álvaro M, Ferrer B, García H, Palomares E, Balzani V, Credi A, Venturi M, Stoddart J, Wenger S (2003) Photochemistry of a dumbbell-shaped multicomponent system hosted inside the mesopores or Al/MCM-41 Aluminosilicate. Generation of long-lived viologen radicals. J Phys Chem B 107:14319–14325.

Anelli P, Spencer N, Stoddart J (1991) A Molecular Shuttle. J Am Chem Soc 113:5131–5133.

Asakawa M, Ashton P, Balzani V, Credi A, Hamers C, Mattersteig G, Montalti M, Shipway A, Spencer N, Stoddart J, Tolley M, Venturi M, White A, Williams D (1998) A Chemically and Electrochemically Switchable [2]Catenane Incorporating a Tetrathiafulvalene Unit. Angew Chem Int Ed Engl 37:333–337.

Ashton P, Ballardini R, Balzani V, Credi A, Dress R, Ishow E, Kleverlann C, Kocian O, Preece J, Spencer N, Stoddart J, Venturi M, Wenger S (2000) A Photochemically-driven Molecular-level Abacus. Chem Eur J 6:3558–3574.

Ashton P, Ballardini R, Balzani V, Baxter I, Credi A, Fyfe M, Gandolfi M, Gomez-Lopez M, Martinez-Diaz M, Piersanti A, Spencer N, Stoddart J, Venturi M, White A, Williams D (1998) Acid-base controllable molecular shuttles. J Am Chem Soc 120:11932–11942.

Astumian R, Hänggi P (2002) Phys Today 55:33.

Aviram A, Ratner MA (1974) Molecular Rectifiers. Chem Phys Lett 29:277.

Bachand G, Soong R, Neves H, Olkhovets A, Craighead H, Montemagno C (2001) Precision Attachment of Individual F1-ATPase Biomolecular Motors on Nanofabricated Substrates. Nano Lett 1:42.

Badjic J, Balzani V, Credi A, Silvi S, Stoddart J (2004) A molecular elevator. Science 303: 1845–1849.

Ballardini R, Balzani V, Credi A, Gandolfi M, Venturi M (2001a) Acc Chem Res 34:445.

Ballardini R, Balzani V, Credi A, Gandolfi M, Venturi M (2001b) Int J Photoenergy 3:63.

Balzani V (2003) Photochem Photobiol Sci 2:479.

Balzani V, Credi A, Venturi M (2003) Molecular Devices and Machines – A Journey into the Nano World. Weinheim: Wiley-VCH.

Balzani V, Credi A, Raymo FM, Stoddart JF (2000a) Artificial Molecular Machines. Angew Chem Int Ed Engl 39:3348.

Balzani V, Mattersteig G, Matthews OA, Raymo FM, Stoddart JF, Venturi M, White AJP, Williams DJ J Org Chem 2000;65:1924 (2000b) Switching of Pseudorotaxanes and Catenanes Incorporating a Tetrathiafulvalene Unit by Redox and Chemical Inputs. J Org Chem 65: 1924–1936.

Balzani V, Clemente-León M, Credi A, Ferrer B, Venturi M, Flood A, Stoddart J (2006) Autonomous Artifical Nanomotor Powered by Sunlight. Proc Natl Acad Sci USA 103:1178–1183.

Balzani V, Scandola F (1991) Supramolecular Photochemistry. Chichester: Horwood.

Bedard T, Moore J (1995) J Am Chem Soc 117:10662.

Bissell A, Córdova E, Kaifer A, Stoddart J (1994) A chemically and electrochemically switchable molecular shuttle. Nature 369:133.

Boyer PD (1993) The binding change mechanism for ATP synthase—Some probabilities and possibilities. Bioch Biophys Acta 1140:215–250.
Brouwer A, Frochot C, Gatti F, Leigh D, Mottier L, Paolucci F, Roffia S, Wurpel G (2001) Photoinduction of Fast, Reversible Translational Motion in a Hydrogen-Bonded Molecular Shuttle. Science 291:2124–2128.
Bustamante C, Keller D, Oster G (2001) The physics of molecular motors. Acc Chem Res 34:412–420.
Carter F (1982) Molecular Electronic Devices. New York: Dekker.
Cavallini M, Biscarini F, Leon S, Zerbetto F, Bottari G, Leigh D (2003) Information Storage using Supramolecular Surface Patterns. Science 299:531
Cecchet F, Rudolf P, Rapino S, Margotti M, Paolucci F, Baggerman J, Brouwer A, Kay E, Wong J, Leigh D (2004) Structural, Electrochemical, and Photophysical Properties of a Molecular Shuttle Attached to an Acid- Terminated Self-Assembled Monolayer. J Phys Chem B 108:15192–15199.
Chen Y, Mao C (2004) Putting a brake on an autonomous DNA nanomotor. J Am Chem Soc 126:8626–8627.
Chen Y, Wang M, Mao C (2004) An autonomous DNA nanomotor powered by a DNA enzyme. Angew Chem Int Ed 43:3554–3557.
Christ T, Kulzer F, Bordat P, Basché T (2001) Watching the photo-oxidation of a single aromatic hydrocarbon molecule Angew Chem Int Ed 40:4192.
Collier CP MG, Wong EW, Luo Y, Beverly K, Sampaio J, Raymo FM, Stoddart JF, Heath JR Science 2000;289:1172 (2000).
Cozzi F, Guenzi A, Johnson C, Mislow K, Hounshell W, Blount J (1981) J Am Chem Soc 103:957.
Dietrich-Buchecker C, Jimenez-Molero M, Sartor V, Sauvage J (2003) Rotaxanes and Catenanes as Prototypes of Molecular Machines and Motors. Pure Appl Chem 75:1383–1393.
Dominguez Z, Khuong T, Dang H, Sanrame C, Nunez J, Garcia-Garibay M (2003) J Am Chem Soc 125:8827.
Drexler E (1986) Engines of Creation The Coming Era of Nanotechnology. New York: Anchor Books.
Drexler E (1992) Molecular Machinery, Manufacturing, and Computation. New York: Wiley.
Elizarov A, Chiu S, Stoddart J (2002) An Acid–Base Switchable [2]Rotaxane. J Org Chem 67:9175–9181.
Endow S, Higuchi H (2000) A mutant of the motor protein kinesin that moves in both directions on microtubules. Nature 406:913–916.
Feynman R (1960a) The Wonders that Await a Micro-Microscope. Sat Rev 43:45–47.
Feynman R (1960b) There's plenty of room at the bottom. Eng Sci Feb 23:22–36.
Finer J, Simmons R, Spudich J (1994) Single myosin molecule mechanics: Piconewton forces and nanometre steps. Nature 368:113–119.
Flood A, Peters A, Vignon S, Steuerman D, Tseng H, Kang S, Heath J, Stoddart J (2004) The Role of Physical Environment on Molecular Electromechanical Switching. Chem Eur J 10:6558–6561.
Forkey J, Quinlan M, Alexander Shaw M, Corrie J, Goldman Y (2003) Three-dimensional structural dynamics of myosin V by single-molecule fluorescence polarization. Nature 422:399–404.
Frey E (2002) Physics in cell biology: on the physics of biopolymers and molecular motors. ChemPhysChem 3:270.
Garaudée S, Silvi S, Venturi M, Credi A, Flood AH, Stoddart JF (2005) Shuttling Dynamics in an Acid-base Switchable [2]Rotaxane. ChemPhysChem 6:2145-2152.
Gimzewski JK, Joachim C (1999) Nanoscale Science of Single Molecules Using Local Probes. Science 283:1683–1688.
Goodsell D (2004) Bionanotechnology: Lessons from Nature. Hoboken: Wiley.
Gust D, Moore T, Moore A (2001) Mimicking Photosynthetic Solar Energy Transduction. Acc Chem Res 34:40–48.
Hernández J, Kay E, Leigh D (2004) A reversible synthetic rotary molecular motor. Science 306:1532–1537.

Hernandez R, Tseng H, Wong J, Stoddart J, Zink J (2004) An operational supramolecular nanovalve. J Am Chem Soc 126:3370–3371.
Hess H, Bachand G, Vogel V (2004) Powering nanodevices with biomolecular motors. Chem Eur J 10:2110–2116.
Hess H, Clemmens J, Howard J, Vogel V (2002) Surface imaging by self-propelled nanoscale probes. Nano Lett 2:113–116.
Hess H, Clemmens J, Quin D, Howard J, Vogel V (2001) Light-Controlled Molecular Shuttles Made from Motor Proteins Carrying Cargo on Engineered Surfaces. Nano Lett 1:235–239.
Hlavina S, Meyer G, Rieder KH (2001) Inducing single-molecule chemical reactions with a UHV-STM: A new dimension for nanoscience and technology. ChemPhysChem 2:361.
Hua W, Chung J, Gelles J (2002) Distinguishing inchworm and hand-over-hand processive kinesin movement by neck rotation measurements. Science 295:844–848.
Huang T, Tseng H, Sha L, Lu W, Brough B, Flood A, Yu B, Celestre P, Chang J, Stoddart J, Ho C (2004) Mechanical Shuttling of Linear Motor-Molecules in Condensed Phases on Solid Substrates. Nano Lett 4:2065–2071.
Itoh H, Takahashi A, Adachi K, Noji H, Yasuda R, Yoshida Y, Kinoshita K (2004) Jr Mechanically driven ATP synthesis by F1-ATPase. Nature 427:465–468.
Jiménez-Molero M, Dietrich-Buchecker C, Sauvage JP (2000) Towards Synthetic Molecular Muscles: Contraction and Stretching of a Linear Rotaxane Dimer. Angew Chem Int Ed 39:3284–3287.
Jiménez-Molero M, Dietrich-Buchecker C, Sauvage JP (2002) Chemically induced contraction and stretching of a linear rotaxane dimer. Chem Eur J 8:1456–1466.
Joachim C, Launay JP (1984) Rotaxane- and catenane-based molecular machines and motors. Nouv J Chem 8:723.
Kaifer A, Gómez-Kaifer M (1999) Supramolecular Electrochemistry. Weinheim: Wiley-VCH.
Katz E, Lioubashevsky O, Willner I (2004) Electromechanics of a redox-active rotaxane in a monolayer assembly on an electrode. J Am Chem Soc 126:15520–15532.
Keaveney C, Leigh D (2004) Shuttling through Anion Recognition. Angew Chem Int Ed 43:1222–1224.
Kelly TR, De Silva H, Silva R (1999) Unidirectional rotary motion in a molecular system. Nature 401:150–152.
Kelly TR, Silva R, De Silva H, Jasmin S, Zhao Y (2000) A Rationally Designed Prototype of a Molecular Motor. J Am Chem Soc 122:6935–6949.
Kern JM, Raehm L, Sauvage JP, Divisia-Blohorn B, Vidal P (2000) Controlled Molecular Motions in Copper-Complexed Rotaxanes: An XAS Study. Inorg Chem 39:1555–1560.
Keyes RW (2001) Fundamental Limits of Silicon Technology. Proc IEEE 89:227.
Kikkawa M, Sablin E, Okada Y, Yajima H, Fletterick R, Hirokawa N (2001) Switch-based mechanism of kinesin motors. Nature 411:439–445.
Koumura N, Geertsema E, Meetsma A, Feringa BL (2000). Light-Driven Molecular Rotor: Unidirectional Rotation Controlled by a Single Stereogenic Center. J Am Chem Soc 122:12005-12006
Koumura N, Zijlstra R, van Delden R, Harada N, Feringa B (1999) Light-driven monodirectional molecular rotor. Nature 401:152–155.
Lehn J (1990) Perspectives in supramolecular chemistry. Angew Chem Int Ed Engl 29:1304.
Lehn J (1995) Supramolecular Chemistry: Concepts and Perspectives. Weinheim: Wiley-VCH.
Lehn J (1996) Comprehensive Supramolecular Chemistry. Oxford: Pergamon Press.
Leigh D, Perez E (2004) Shuttling through reversible covalent chemistry. Chem Commun 20:2262–2263.
Leigh D, Wong J, Dehez F, Zerbetto F (2003) Unidirectional rotation in a mechanically interlocked molecular rotor. Nature 424:174–179.
Li J, Tan W (2002) A single DNA molecule nanomotor. Nano Lett 2:315–318.
Limberis L, Stewart R (2000) Toward kinesin-powered microdevices. Nanotechnology 11:47–51.
Long B, Nikitin K, Fitzmaurice D (2003) Self-assembly of a tripodal pseudorotaxane on the surface of a titanium dioxide nanoparticle. J Am Chem Soc 125:5152–5160.

Luo Y, Collier C, Jeppesen J, Nielsen K, Delonno E, Ho G, Perkins J, Tseng H, Yamamoto T, Stoddart J, Heath JC (2002) Two-dimensional molecular electronics circuits. ChemPhysChem 3:519–525.

Mao C, Sun W, Shen Z, Seeman N (1999) A nanomechanical device based on the B-Z transition of DNA. Nature 397:144–146.

Marcaccio M, Paolucci F, Roffia S (2004) Trends in Molecular Electrochemistry. New York: Dekker.

Mehta AD, Rief M, Spudich J, Smith D, Simmons R (1999) Single-molecule biomechanics with optical methods. Science 283:1689–1695.

Metzger R (2003) Unimolecular electrical rectifiers. Chem Rev 103:3803–3834.

Mobian P, Kern J, Sauvage J (2004) Light-Driven Machine Prototypes Based on Dissociative Excited States: Photoinduced Decoordination and Thermal Recoordination of a Ring in a Ruthenium(ii)-Containing [2]Catenane. Angew Chem Int Ed 43:2392–2395.

Moerner W (2002) A Dozen Years of Single-Molecule Spectroscopy in Physics, Chemistry, and Biophysics. J Phys Chem B 106:910.

Molloy J, Veigel C (2003) Myosin motors walk the walk. Science 300:2045–2046.

Montemagno C, Bachand G (1999) Constructing nanomechanical devices powered by biomolecular motors. Nanotechnology 10:225.

Niemeyer C (1997) DNA as a Material for Nanotechnology. Angew Chem Int Ed 36:585–587.

Niemeyer C (2000) Self-assembled nanostructures based on DNA: Towards the development of nanobiotechnology. Curr Opin Chem Biol, 4:609–618.

Noji H, Yasuda R, Yoshida M, Kinosita KJ (1997) Direct observation of the rotation of the F1 -ATPase. Nature 386:299–302.

Oster G, Wang H (2000) Reverse engineering a protein: the. • mechanochemistry of ATP synthase. Bioch Biophys Acta 1458:482.

Parisi G (2005) Nature 433:221.

Poleschak I, Kern JM, Sauvage JP (2004) A copper-complexed rotaxane in motion: pirouetting of the ring on the millisecond timescale. Chem Commun:474.

Raehm L, Kern JM, Sauvage JP (1999) A Transition Metal Containing Rotaxane in Motion: Electrochemically Induced Pirouetting of the Ring on the Threaded Dumbbell. Chem Eur J 5:3310.

Rastogi V, Girvin M (1999) Structural changes linked to proton translocation by subunit c of the ATP Synthase. Nature 402:263.

Reconditi M, Linari M, Lucii L, Stewart A, Sun Y-B, Boesecke P, Narayanan T, Fischetti RF, Irving T, Piazzesi G, Irving M, Lombardi V (2004) The myosin motor in muscle generates a smaller and slower working stroke at higher loads. Nature 428:578–581.

Ren H, Allison W (2000) On what makes the gamma subunit spin during ATP hydrolysis by F(1). Bioch Biophys Acta 1458:221.

Rigler R, Orrit M, Talence I, Basché T (2001) Single Molecule Spectroscopy. Berlin: Springer Verlag.

Rondelez Y, Tresset G, Nakashima T, Kato-Yamada Y, Fujita H, Takeuchi S, Noji H (2005) Highly coupled ATP synthesis by F1-ATPase single molecules. Nature 433:773.

Sambongi Y, Iko Y, Tanabe M, Omote H, Iwamoto-Kihara A, Ueda I, Yanagida T, Wada Y, Futai M (1999) Rotation of a Complex of the $\gamma$ Subunit and c Ring of Escherichia coli ATP Synthase. Science 286:1722.

Samori B, Zuccheri G, Baschieri P (2004) Protein Unfolding and Refolding Under Force: Methodologies for Nanomechanics. ChemPhysChem 6:29–34.

Sauvage JP, Dietrich-Buchecker C (1999) Molecular Catenanes, Rotaxanes and Knots. Weinheim: Wiley-VCH.

Schliwa M, Woehlke G (2003) Molecular Motors. Nature 422:759–765.

Seelert H, Poetsch A, Dencher N, Engel A, Stahlberg H, Müller D (2000) Structural biology: Proton-powered turbine of a plant motor. Nature 405:418–419.

Sherman W, Seeman N (2004) A precisely controlled DNA biped walking device. Nano Lett 4:1203–1207.

Shima T, Hampel F, Gladysz J (2004) Molecular Gyroscopes: $\{Fe(CO)^{3}\}$ and $\{Fe(CO)^{2}(NO)\}+$ Rotators Encased in Three-Spoke Stators; Facile Assembly by Alkene Metatheses. Angew Chem Int Ed 43:5537.

Shinkai S, Ikeda M, Sugasaki A, Takeuchi M (2001) Positive Allosteric Systems Designed on Dynamic Supramolecular Scaffolds: Toward Switching and Amplification of Guest Affinity and Selectivity. Acc Chem Res 34:494–503.

Smalley R (2001) "Of chemistry, love and nanobots - How soon will we see the nanometer-scale robots envisaged by K. Eric Drexler and other molecular nanotechologists? The simple answer is never". Sci Am 285:76–77.

Soong R, Bachand G, Neves H, Olkhovets A, Craighead H, Montemagno C (2000) Powering an inorganic nanodevice with a biomolecular motor. Science 290:1555.

Steed J, Atwood J (2000) Supramolecular Chemistry. New York: Wiley.

Steed J, Atwood J (2004) Enclyclopedia of Supramolecular Chemistry. New York: Dekker.

Steinberg-Yfrach G, Liddell P, Hung S, Moore A, Gust D, Moore T (1997) Conversion of light enery to proton potential in liposomes by artificial photosynthetic reaction centers. Nature 385:239–241.

Steinberg-Yfrach G, Rigaud J, Durantini E, Moore A, Gust D, Moore T (1998) Lightdriven production of ATP catalysed by F0F1- ATP synthase in an artificial photosynthetic membrane. Nature 392:479.

Stock D, Leslie A, Walker J (1999) Molecular Architecture of the Rotary Motor in ATP Synthase. Science 286:700.

Svoboda K, Schmidt C, Schnapp B, Block S (1993) Direct observation of kinesin stepping by optical trapping interferometry. Nature 365.

Ter Wiel MKJ, Van Delden RA, Meetsma A, Feringa BL (2003) Increased Speed of Rotation for the Smallest Light-Driven Molecular Motor. J Am Chem Soc 125:15076–15086.

Tseng H, Vignon S, Stoddart J (2003) Toward Chemically Controlled Nanoscale Molecular Machinery. Angew Chem Int Ed 42:1491–1495.

Tseng H, Vignon S, Celestre P, Perkins J, Jeppesen J, Di Fabio A, Ballardini R, Gandolfi M, Venturi M, Balzani V, Stoddart J (2004) Redox-Controllable Amphiphilic [2]Rotaxanes. Chem Eur J 10:155-172.

Vale R, Milligan R (2000) The way things move: Looking under the hood of molecular motor proteins. Science 288:89–95.

Van Delden R, Koumura N, Schoevaars A, Meetsma A, Feringa B (2003) A donor–acceptor substituted molecular motor: Unidirectional rotation driven by visible light. Org Biomol Chem 1:33–35.

Visscher K, Schnitzer M, Block SN (1999) Single kinesin molecules studied with a molecular force clamp. Nature 400:184–189.

Walker J (1998) ATP Synthesis by Rotary Catalysis (Nobel lecture). Angew Chem Int Ed Engl 37:2308.

Wang Q, Qu D, Ren J, Chen K, Tian H (2004) A lockable light-driven molecular shuttle with a flourescent signal. Angew Chem Int Ed 43:2661.

Willner I, Pardo-Yssar V, Katz E, Ranjit K (2001) A photoactivated 'molecular train' for optoelectronic applications: light-stimulated translocation of a beta.-cyclodextrin receptor within a stoppered azobenzene-alkyl chain supramolecular monolayer assembly on a Au-electrode. J Electroanal Chem 497:172–177.

Xi J, Schmidt J, Montemagno C (2005) Self-assembled microdevices driven by muscle. Nature Mater 4:180–184.

Yan H, Zhang X, Shen Z, Seeman N (2002) A robust DNA mechanical device controlled by hybridization topology. Nature 415:62–65.

Yasuda R, Noji H, Kinosita KJ, Yoshida M (1998) F1-ATPase Is a Highly Efficient Molecular Motor that Rotates with Discrete 120° Steps. Cell 93:1117–1124.

Yasuda R, Noji H, Yoshida M, Kinosita K, Itoh HN (2001) Resolution of distinct rotational substeps by submillisecond kinetic analysis of F1-ATPase. Nature 210:898–904.

Yidiz A, Tomishige M, Vale R, Selvin P (2004) Kinesin walks hand-over-hand. Science 303:676–678.

Yildiz A, Forkey J, McKinney S, Ha T, Goldman Y, Selvin P (2003) Myosin V walks hand-over-hand: Single fluorophore imaging with 1.5-nm localization. Science 300:2061–2065.
Yin P, Yan H, Daniell X, Turberfield A, Reif J (2004) A unidirectional DNA walker that moves autonomously along a track. Angew Chem Int Ed 43:4906–4911.
Yurke B, Turberfield A, Mills Jr A, Simmel F, Neumann J (2000) A DNA- fuelled molecular machine made of DNA. Nature 406:605.
Zander C, Enderlein J, Keller R (2002) Single Molecule Detection in Solution. Weinheim Wiley-VCH.
Zheng X, Mulcahy M, Horinek D, Galeotti F, Magnera T, Michl J (2004) Dipolar and Nonpolar Altitudinal Molecular Rotors Mounted on an Au(111) Surface. J Am Chem Soc 126: 4540–0215.

# Micro and Nano Engineered Extracellular Matrices

**James J. Norman and Tejal A. Desai**

## 1 Introduction

Cells in our body are in intimate contact with the extracellular matrix (ECM), the substratum in which cells live. The ECM is composed of proteins, glycoproteins and proteoglycans that are arranged in tissue-specific structures. Cells form adhesions to the ECM via integrins, transmembrane proteins that interact with specific amino acid sequences found within the proteins that make up the ECM (for example, the arginine-glycine-aspartic acid, RGD, sequence). These adhesions are one way the cell interacts with its environment. The cell can influence the physical arrangement of the ECM through these adhesions and the cell itself can be influenced through the adhesions. When cells migrate, they extend psuedopods (filipodia) from the main body of the cell. Gustafson and Wolpert have made striking observations of psuedopods on the order of 500 nm extending from the body of a cell (sea urchin mesenchymal cells) to explore and probe the surrounding environment (Gustafson and Wolpert 1999). This exploration by the cells appears to happen randomly with psuedopods sweeping the surface until a point of stable contact is made. Cell movement then occurs in the direction of the contact through retraction of the attached psuedopod. This apparent exploration by cells of their surroundings has led researchers to develop scaffolds that provide cues to the cells as they migrate across its surface. Finding the exact cues that will cause the cells to make a

J.J. Norman
Warner Babcock Institute for Green Chemistry, LLC,
Wilmington, MA, USA

T.A. Desai (✉)
Department of Physiology and Division of Bioengineering, University of California,
San Francisco, CA, USA
e-mail: tejal.desai@ucsf.edu

G.A. Silva and V. Parpura (eds.), *Nanotechnology for Biology and Medicine: At the Building Block Level*, Fundamental Biomedical Technologies,
DOI 10.1007/978-0-387-31296-5_5, 

preferential adhesion to the surface has been one of the major endeavors for tissue engineers over the last decade. Chemical and physical cues are the two most widely studied modifications that are made to the surface of a material. One of the most difficult tasks is determining the role played by each of these treatments and finding the proper balance between them for optimal cell attachment and growth. Physical cues, artificially created on the surface of a material, alone have been shown to significantly affect organization of cells in culture through the phenomenon of contact guidance (Hanarp et al. 1999). This has been demonstrated in both two-dimensional and three-dimensional cultures with artificial tissues being created that are strikingly similar in both appearance and function to natural tissues (Deutsch et al. 2000; Norman and Desai 2005). The key to recapitulating tissue organization and function in vitro may lie in the ability to recreate the cells' ECM environment.

The ECM has considerable topographic detail at the nanometer scale. It has been very well established that micrometer-size features on a culture surface can affect parameters such as migration, adhesion and morphology independent of biochemical cues presented to the cells (Rahul Singhvi 1994). Further evidence of topographic control of cellular growth, independent of biochemical modulation, was shown by growing bovine aortic endothelial cells on polymer casts of the subendothelial ECM (Goodman et al. 1996). Polyurethane casts of the subendothelial ECM of arteries and veins showed fidelity of feature replication down to 50 nm. After several days of culture on the cast, cells appeared more similar to those in native arteries than cells cultured on untextured surfaces. Obviously, the nanometer-size features of the ECM alone can go a far way to recreating in vivo cellular characteristics.

As tissue engineers, all size scales of a scaffold must be considered when engineering an artificial tissue: the superstructure, the overall shape of the scaffold; the microstructure, the cellular-level structure of the surface; and the nanostructure, the subcellular-level structure of the surface. The first two levels of this hierarchy have been very well studied over the years and methods for controlling them are well established. The nanoscale structure and its influence on cells, however, is less well understood, yet just as important as the levels above it. When cells are cultured outside the body on a polymeric scaffold, or any one of a variety of materials used as substrates for a biomedical device or implant, often, only the macro- and microscale features are considered. However, the nanoscale features of all of these materials may greatly affect, for better or worse, the culture in question. The inherent nanotopography encountered by the cells on such materials is most likely different from that found in their native environment and may provide physical cues that lead to less than desirable growth and function.

Understanding the architectural make-up of the ECM and how it influences the cells in contact with it will allow scientists and engineers to design materials that mimic what the cells "see" when they explore their native environment. Researchers are now looking to see if and how domains of nanometer-size variations in topography of the ECM affect the cells growing on them. Investigative studies into the nanoscale topography of basement membranes from different species (including human) and different tissues have shown varied and rich nanoscale topography with features ranging from only a few to hundreds of nanometers (Yamasaki et al. 1990; Shirato et al. 1991; Kubosawa and Kondo 1994; Abrams et al. 2000a, b, 2002, 2003). A variety of methods have been used to study tissue samples including atomic force microscopy (AFM), transmission electron microscopy (TEM) and scanning electron microscopy (SEM). The basement membranes were found to be a composed of a complex arrangement of fibers and pores that result in a landscape rich in three-dimensional features. These features match the size scale of a cell's processes that explore its environment and the adhesion sites it makes with the ECM. Modeling of the surface of the macaque corneal epithelium basement membrane preformed by Abrams et al. shows an amplification of surface area for cell interactions of 400% over flat culture surfaces (Abrams et al. 2000a). This large increase in surface area for cell–matrix interactions to take place could have a large influence on the abilities of cells to attach and migrate to the surface of the ECM.

The use of AFM has allowed for the quantification of feature height, frequency and overall roughness of the basement membrane. This is important because many micro and nanofabrication techniques are designed to create features of specified height and frequency. With many techniques it is now possible to recreate the cells natural environment at a scale equivalent to that which an individual cell would experience. For example, chemical etching of materials to control the nanoscale roughness of a surface is a technique that is being widely investigated for studying nanotopography and its effects on cellular behavior. These studies have led to the engineering of surfaces for in vitro culture that display the same type of nanoscale topography to cells as found in vivo (Fan et al. 2002; Thapa et al. 2003; Pattison et al. 2005).

What type of topographical features to create and the technique used to create them will depend on the tissue type and what aspect of cellular growth needs to be modulated. Determining the appropriate factors will involve a detailed study of the physical environment of the tissue in question as well as creating model surfaces to study the effects on living cells. The types of nanotopographical features that can be created on materials fall into two main categories: unordered topographies and ordered topographies. Unordered topographies are typically those which spontaneously occur during processing. Examples of such topographies can be made using techniques such as polymer demixing, colloidal lithography and chemical etching. Surfaces patterned in this manner tend to give features that are random in orientation and organization with imprecise or no control over feature geometry. These techniques, however, are usually simpler, quicker and less costly than the more complex equipment and processes needed to create ordered topographies. Ordered topographies are those which can be created with techniques like photolithography and electron beam lithography. These methods allow the creation of prescribed patterns that are well ordered and geometrically precise. However, they usually require very expensive equipment and a high level of expertise. The value of an ordered topography over an unordered topography was investigated by Curtis et al. (2001). In their experiment, surfaces of various nanoscale ordered patterns were created using electron beam lithography, and surfaces with nanoscale unordered patterns were created using colloidal lithography. Rat epitonon fibroblasts showed higher level of adhesion to the surface with an unordered pattern than to that of both planar surfaces and ordered surfaces. The surfaces with the ordered patterns had even lower levels of adhesion than flat surfaces. It is clear from studies such as this that not only must size and shape of the features be considered by the bioengineer when designing a scaffold but also the technology used must be carefully considered.

What follows is a review of several applications of engineered cellular environments that are designed to either recreate native tissue architecture or control the environment of cells so they may better survive in vivo transplantation. In all of these cases, the engineering of the environment is used to allow cells to grow and function in a manner that they would in vivo. Micro- and nanotechnology play an important role in the fabrication of all of these environments. The techniques employed either provide the cells with a natural ECM environment so the cells can interact with it yet control the organization of this environment to force cells to take on natural tissue patterns, or they create nanoscale conduits (pores) to control the movement of molecules in and out of the cells' environment.

## 2 Nanoporous Alumina Membranes for Bone Tissue Engineering

The success of orthopedic implants depends on the native bone tissue making and retaining stable fixation to the implant. A better understanding of the events that take place at the bone–material interface is needed in order to engineer materials that will promote an interfacial layer between tissue and implant with adequate healing and biomechanical properties. Materials that promote osseointegration can help speed the healing process and strengthen the bone–implant interface.

Alumina, a synthetically produced amphoteric aluminum oxide $Al_2O_3$, is a scaffold material that is biocompatible and has current applications in orthopedic and dental implants. Nanoscale features in bone have led to taking materials such as alumina and patterning them with features of the same size scale. It is hoped that this may lead to better integration into the host bone as well as improved biocompatibility. Nanoporous alumina has received much attention for its use in bone tissue engineering. The use of nanoporous alumina stems from the idea that the feature size matches that of the inorganic particles in bone. Swan, et al. have produced porous alumina membranes with pore sizes ranging from 30 to 80 nm created using a two-step anodization process (Fig. 1) (Swan et al. 2005a). These membranes were fabricated from aluminum sheeting with a process adapted from Gong et al. (2003). The aluminum is coated on one side with butyl acetate. This acts as a protective barrier during the anodization process (Fig. 1a). The first step in making the pores is the creation of dimples on the uncoated surface of the aluminum. Anodization occurs in a glass chamber where the aluminum sheet acts as the anode and platinum foil acts as the cathode. Both the aluminum sheet and the platinum are suspended from electronic leads into a bath of 0.25 M oxalic acid, which acts as the electrolyte. This step creates an oxide layer on the aluminum surface. The resultant pore size can be

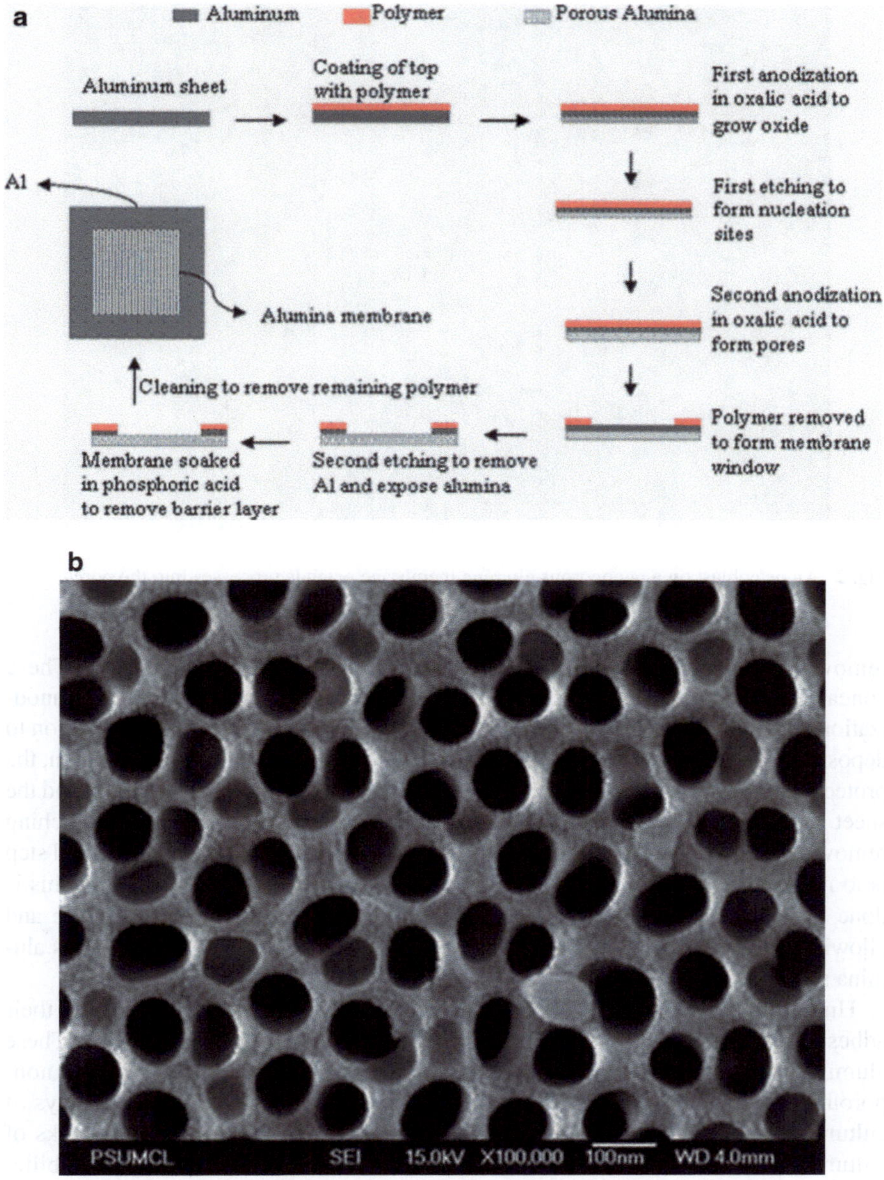


**Fig. 1** (**a**) Flow diagram of the steps involved in creating nanoporous alumina membranes. (**b**) SEM of a nanoporous alumina scaffold. Average pore size is 89 ± 12 nm. Anodization voltage was 60 V

controlled by controlling the DC-voltage applied to the leads. There is a theoretical linear relationship between the voltage and the pore diameter of 1.29 V/nm. The next step in the process involves etching the aluminum sheet in a solution of 4% chromic acid and 8% phosphoric acid. The sheet is etched until the entire oxide layer is

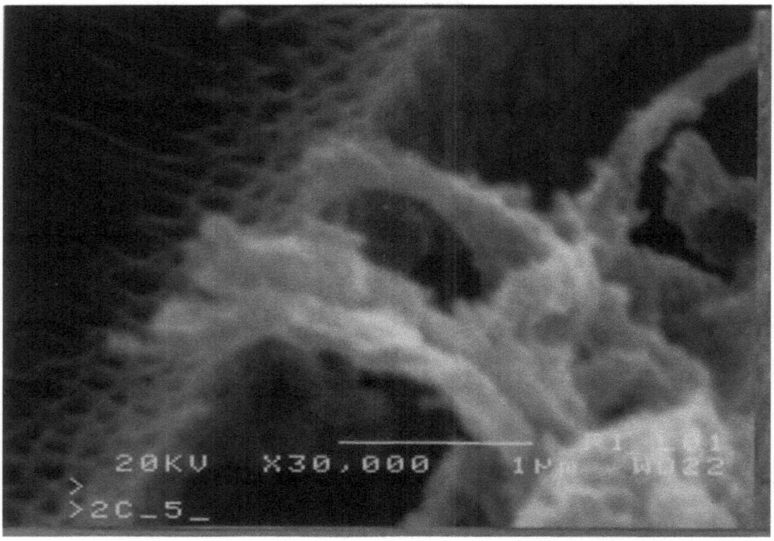


**Fig. 2** An osteoblast on a nanoporous alumina membrane extends processes into the pores

removed. This process creates a periodic concave pattern on the surface. These concavities act as the nucleation sites for the pores that are created in the second anodization and etching step. The second anodization is a repeat of the first anodization to deposit an even layer of alumina on the surface of the sheet. After this anodization, the protective polymer layer is stripped from the backside to expose the aluminum and the sheet is soaked in a mixture of copper chloride and hydrochloric acid. The etching removes the exposed aluminum and exposes the alumina layer beneath. The final step is to remove the alumina layer deposited in the second anodization process. This is done by placing a few drops of 10% phosphoric acid on the alumina surface and allowing it to soak until the pores are exposed. The SEM image of a nonporous alumina scaffold produced using the above outline process is shown in Fig. 1b.

Human fetal osteoblasts were cultured on the nanoporous membranes and their adhesion, proliferation, performance and ECM production were monitored. These alumina nanoporous scaffolds were found to increase osteoblast adhesion over nonporous membranes. Adhesion and proliferation were increased for up to 4 days of culture (Popat et al. 2005) while matrix production was increased for 4 weeks of culture. Further modification of the porous membranes was achieved by immobilizing RGDC moieties onto the alumina (Swan et al. 2005b). The addition of this peptide sequence did not clog the pores and increased osteoblast adhesion as well as matrix production 2 days into culture. The introduction of nanopores changes what the cells sense as they grow on the surface. How individual cells are interacting with the nanoporous membrane is of particular importance. Scanning electron microscopy of osteoblasts growing on the nanoporous membranes shows that cells extend processes into the nanopores (Fig. 2). The cells appear to be exploring their environment with particular interest in the pores. It is possible the cells are trying to migrate into the material through extending filiopdia into the pores or the pores provide a more attractive anchoring site than the flat surfaces of the membrane.

Taken together, these results suggest that the osteoblasts are reacting to the nanopores by anchoring themselves faster and more strongly to the membranes that have pores. The pores themselves are at a scale comparable to cellular extensions, and the in growth of extensions into the membrane pores suggest that osseointegration could be enhanced by creating nanopores on prosthetic implant surfaces.

## 3 Cardiac Tissue Engineering on Microfabricated Membranes

One of the major challenges in engineering tissues is mimicking the complex cellular organizations and functions of the native tissues of the human body. Tissue structure and function are very highly interrelated in most cases. The cellular and macromolecular organization of the tissue often brings about mechanical and biological functionality. For example, it is the circumferential arrangement of smooth muscle fiber layers that allows for change in the caliber of the lumen of blood vessels (Fawcett 1986); the wickerwork pattern of collagen fibers in the skin give it mechanical strength (Alberts et al. 1994); and the end-to-end connections and parallel arrangement of myocytes allow the ventricle to contract as a unit and eject blood (Sommer 1995). Without proper cellular organization, an artificial tissue does not function adequately.

In order to recapitulate cardiac myocyte structure and function in vitro, microtextured silicone substrates were designed to induce the natural end-to-end, parallel arrangement of myocytes. The substrates were designed to have rows of micropegs (5 μm high), parallel groves (5 μm wide) or a combination pattern where the pegs were spaced within the grooves. These cardiac tissue-tailored substrates were produced through a series of photolithography and microfabrication techniques. Neonatal rat ventricular myocytes (NRVM) were plated on these substrates and compared to those on nontextured substrates. NRVM displayed an increase in myofibrillar height and decrease in cell area that did not affect the stoichiometry of the myofibrillar proteins. The microtopography did elicit microenvironmental remodeling of proteins that mechanically attach the cell to its surroundings. Cells on nontextured surfaces ended in long non-striated cables while those on pegged substrates ended in sarcomeric striations at the vertical face of the pegs. Expression of focal adhesion proteins, such as vinculin and paxillin, were decreased on combination surfaces as compared to nontextured substrates (Motlagh et al. 2003b). Through the use of the microgrooved surface topography, it was possible to orient the cardiac myocytes in vitro. The addition of microgrooves increased the end-to-end cell–cell contact and the expression of both N-cadherin (mechanical) and connexin43 (electrical) junctions reminiscent of appearance in the intact neonatal heart (Fig. 3) (Motlagh et al. 2003a).

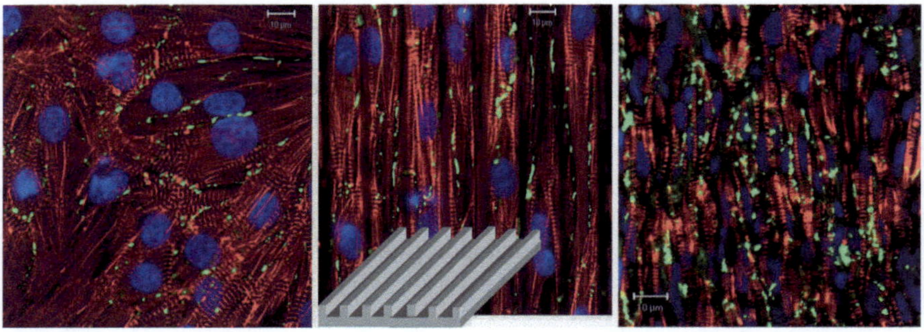

**Fig. 3** Orientation of neonatal cardiac myocytes grown on grooved surfaces mimic that found in intact neonatal heart. *Left*, growth on non-textured surface. *Center*, growth on grooved surface; *corner inset* presents a three-dimesional diagram of a grooved substrate. *Right*, frozen section from an intact neonatal heart. Confocal fluorescence images show electrical gap junction connexin43 protein (*green*), actin filaments (*red*) and nuclei (*blue*)

One important area of cardiac tissue engineering is the control of fibroplasia (excessive fibroblast growth) in primary cultures of cardiac myocytes. Neonatal rat ventricular fibroblasts (NRVF) were cultured on the microtextured surfaces to determine how the topography of the substrate influences the fibroblast population. Cell proliferation decreased markedly on microtextured membranes as measured by cell counts and Western blotting of cyclin D1. Myocyte to fibroblast ratios are higher on microprojections, due to reduced proliferation of fibroblasts (Boateng et al. 2003). It was also determined that the mechanisms underlying these responses depend upon focal adhesion stabilization by use of a Rho kinase inhibitor (Y27632) to decrease NRVF mobility. It is hypothesized that cells sense the reactive forces on attachment to the micropeg. These forces are transferred from the membrane to the cytoskeleton, which leads to blockage of proliferation, which in turn matures them.

By examining the organizational pattern of the ventricular mycoytes in vivo, a substrate was fabricated that was tailored to control cellular and subcellular organization in vitro. The cells were able to recognize variations in the substrate architecture and to internalize these forces localized to small parts of the cell membrane in order to influence protein organization and to effect genetic expression leading to more natural growth.

## 4 Three-Dimensional Scaffolds for Tissue Engineering

The importance of culturing cells in three-dimensional (3D) artificial matrices to mimic the body's own ECM is becoming more apparent. The third dimension has numerous advantages for cells (Powell 2005). Cells show a more life-like gene expression profile and subcellular architecture when grown in three-dimensional instead of conventional flat two-dimensional (2D) surfaces (Cukierman et al. 2001; Pedrotty et al. 2005; Powell 2005).

Three-dimensional matrices have also provided new insight into tissue organization and more accurate models for pathogenesis than traditional 2D culture techniques, such as Petri dishes. Work done studying human breast cancer has shown the importance of moving from traditional, nonphysiological 2D systems to 3D ones. Culture of human breast epithelial cells in collagen gels has demonstrated the ability of the cells to re-express their in vivo organization and differentiation, and recapitulate histology. These studies also allowed for the identification of previously unknown phenomena at many different levels; cell–cell interactions, gene expression and ECM affects on cellular organization and polarity which were not present in traditional 2D cultures (Bissell et al. 2002, 2003; Gudjonsson et al. 2003). Three-dimensional gel culture systems also facilitate natural, complex 3D tissue development in vivo. Co-cultures of human umbilical-vein epithelial cells and 10T1/2 mesenchymal cells in collagen-fibronectin gels when implanted into mice developed a 3D branching network of tubes that connected with the mouse's own vasculature and became perfused (Koike et al. 2004). Arteriolar and venular patterns of blood flow were observed in the construct and the new vessels responded naturally to local administration of a vasoconstrictor. The use of 3D gels has also been shown to provide an improved environment for differentiation of embryonic stem cells. Chondrogenic differentiation of embryoid bodies in polyethylene glycol (PEG) based gels in comparison to monolayer cultures showed induction of chondrocytic phenotype and upregulation of cartilage relevant markers (Alhadlaq and Mao 2004). Histological analysis demonstrated basophilic ECM deposition characteristic of neocartilage (Hwang et al. 2005). It is clear that providing cells with the cues present in a 3D environment is essential to creating complex engineered tissue systems.

The most common way of creating a 3D cell culture is to suspend cells in Type I collagen. This provides not only physical support but also a natural biochemical and physical environment for the cells to interact with. Other natural gels such as gelatin and Matrigel (a commercially available basement membrane) or synthetic gels such as PEG are also commonly used. These gels can provide a nanotopgraphy to cells that is similar to what they would naturally find in vivo (Abrams et al. 2000a). This has the advantage of influencing individual cells to grow in a more natural manner. Unfortunately, cellular organization in three dimensions of cell populations is not possible with these traditional 3D scaffolds because many of the factors influencing tissue morphogenesis are still unknown, and in an isolated system, it is not possible to recreate the entire dynamic physical and biochemical milieu with which a developing or healing organ would be supplied. However, if an engineered ECM can be provided with natural details at the scale with which individual cells can interact, with the ECM still properly ordered at a larger scale, it may be possible to build a tissue or organ in the proper manner without going through the entire developmental or healing process.

The traditional manner of seeding cells into a 3D gel is to mix the cells into a prepolymer solution in order to disperse them through the matrix. Then the polymer is cured to entrap the cells and provide form to the construct. Alternatively, the cells may be placed on the surface of a cured polymer matrix into which they can then migrate. These methods of seeding, however, provide little to no control over the resultant cellular organization. Therefore, a technique is needed to move from

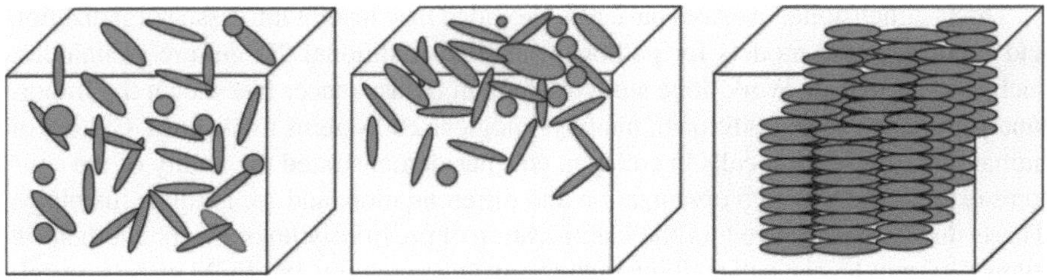

**Fig. 4** Schematic showing the lack of control provided by current scaffolding techniques and the type of control needed for true tissue engineering. (*Left*) Cells are mixed into the matrix or (*Center*) cells are seeded on top of the matrix. Neither of these approaches provides control over cellular organization. (*Right*) Cells show a precise, controlled 3D organization in the matrix. This is not possible with current scaffolding techniques. Cells are represented by *gray ovals*

creating unorganized groups of cells to a 3D-engineered tissue with a high degree of organization at the level of individual cells (Fig. 4). A scaffold that could induce proper cellular organization in the x, y and z directions would allow for the construction of artificial tissues that recapitulate the properties of the natural counterpart.

It has been shown that arrays of microgrooves on culture surfaces can organize monolayers of cells in the direction of the pattern. This was accomplished by taking advantage of the phenomena of contact guidance that has been shown to provide control over the organization of cells in two-dimensional cultures (Dunn and Heath 1976). Contact guidance has been demonstrated to have an influence with many different cell types including fibroblasts (den Braber et al. 1998), cardiac myocytes (Deutsch et al. 2000), smooth muscle cells (Sarkar et al. 2005), neurons (Degenaar et al. 2001) and bone marrow cells (Matsuzaka et al. 2003). All of these studies have shown that surface topography plays an important role in cellular organization. The texture of the surface has also been shown to have significant influence on fibroblast adhesion and proliferation (van Kooten et al. 1998). These observations led us to design an extension of this system that would provide similar control over organization in three dimensions. A three-dimensional artificial tissue system was created that has organization at the cellular level similar to that which would be found in a native in vivo system.

The design of the scaffold is a microfabricated 3D array of deep, parallel channels that is integrated into the interior of a collagen matrix. The ordered pattern of polydimethylsiloxane (PDMS) acts as a skeleton to guide the cells growing in the collagen matrix (Fig. 5). The channels are created in PDMS using standard photolithography and soft lithography techniques. A collagen gel seeded with fibroblasts is then molded around the PDMS skeleton and the entire construct is cultured under standard cell culture conditions. By using the natural collagen matrix to suspend the cells, the cells are provided with an ECM environment that is similar to what would be found in the body while the PDMS internal skeleton provides a physical barrier that the cells respond to by organizing in relation to the channels.

Cellular parameters varied significantly between control gels and scaffolds. Of most interest was the orientation of the cells in relation to the scaffold. Control gels,

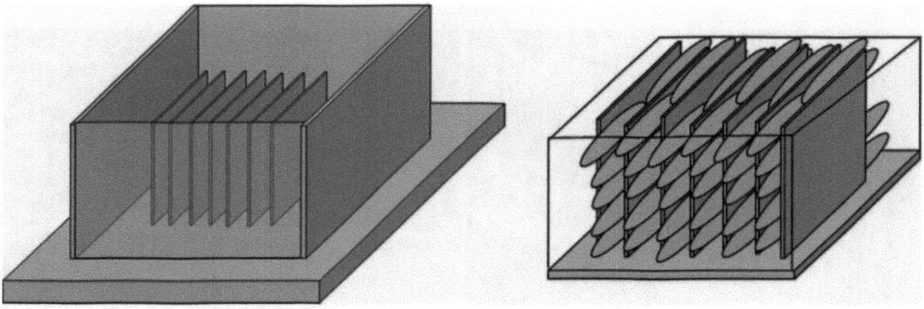

**Fig. 5** Schematic showing desired effects of using an internal scaffold to control three-dimensional cellular organization in a matrix. (*Left*) The PDMS scaffold design. (*Right*) The pattern of cellular organization created using this design

which contain no internal skeleton to provide contact guidance, allowed cells to grow randomly throughout the gel. No underlying organizational pattern is observed. With the addition of an internal scaffold to control cell migration, the organization of the cells becomes strikingly clear. The cells were shown to be highly aligned with the pattern of the scaffold at every level in which they appeared within the three-dimensional construct as seen by confocal fluorescence microscopy (Fig. 6b). While this system was used as a model for making a composite scaffold that can guide cell growth, it does have physiological relevance to tissue systems such as tendon and muscle which have similar cellular organization. In addition, cell morphology was found to differ in the scaffolds, where cells were longer and thinner than those grown in control gels. This is presumed to be because of their interaction with the PDMS skeleton. It is still unclear how the cells are interacting with the PDMS walls of the scaffold. They may be treating it as a surface and growing along it as if they were on a flat, 2D culture turned on its side. The reduction in the observed minor axis and the increased spread in the major axis may be indicative of the cell flattening and spreading on the wall. In this case, this skeleton could provide further surface area for nanoscale modification for the cells to interact with.

As it is, the fusion of a scaffold with a precisely defined architecture with a matrix that can support cell growth led to an engineered construct that takes on a three-dimensional organization not possible with standard 3D culture techniques. Further work needs to be done in order to assess how differing scaffold and matrix parameters affect the cells. One area of particular interest is the use of biodegradable polymers to replace the PDMS of the scaffold. PDMS serves as a model system, but once the cells proliferate and organize, it would be desirable to remove the internal scaffold. This could be accomplished with the use of a biodegradable polymer designed to degrade at some point after the cells organize. Also, the incorporation of pores into the PDMS skeleton could allow better diffusion of nutrients to cells. This should be done without disrupting the architecture of the pattern itself as to not lose the control over organization provided. Methods for accomplishing this are addressed in the next segment.

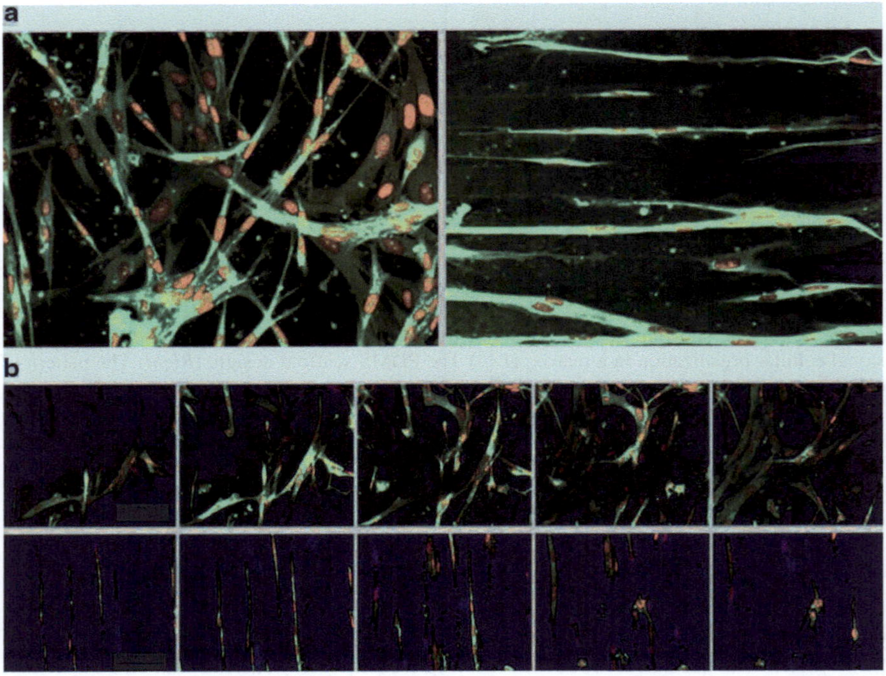


**Fig. 6** (**a**) Random orientation of cells grown in a gel without scaffolding capability (control gel, *left*) and their patterned orientation when grown in a scaffolded gel (*right*). (**b**) Confocal fluorescence microscopy *z*-sections (1 μm apart) though a portion of a control (*top*) and scaffolded (*bottom*) gel

Success with controlling the organization of individual cells within a three-dimensional population could lead to the engineering of superstructures of cells that more closely resemble those of the human body. The PDMS/collagen gel composite technique we have employed, however, can allow for complex three-dimensional patterns to be created for engineering virtually any tissue where organization of the cells relative to each other is important.

Another very active area of three-dimensional tissue engineering research is in the development of a tissue-engineered, small-diameter (<6 mm) vascular graft. One of the most difficult aspects of the vasculature to recreate in biological constructs has been the strength of native vessels to withstand the physiological pressures. The medial layer of blood vessel is composed of multiple layers of smooth muscle cells arranged in alternating spiral layers (Rhodin 1980). The organization of ECM proteins, such as collagen and fibronectin and elastin, also play an important role in vessel strength and integrity. The medial layer provides the strength, elasticity and contractility to the vessel. The patterning of the medial layer is one aspect that has never been investigated in artificial constructs. It is believed that this distinct architecture plays a role in the function of the medial layer of the vessel.

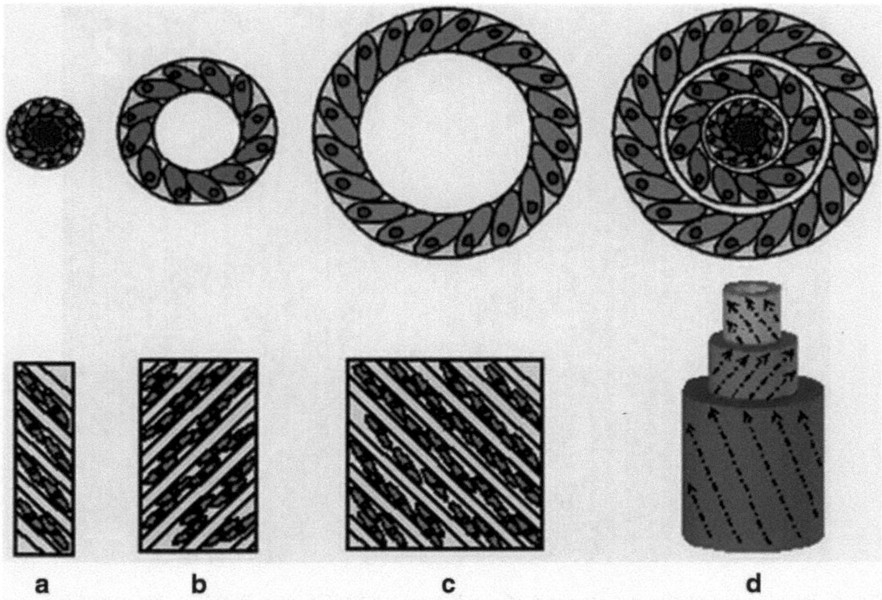


**Fig. 7** A schematic of the smooth muscle cell organization in the medial layer of elastic arteries as proposed by Rhodin (1980). Each layer (**a–c** shows an end on view of the vessel) is arranged in a spiral pattern (*top*) that if unrolled would show the cells organized in parallel strips (*bottom*). By layering the sheets and rolling into a cylinder, the herringbone pattern of vascular smooth muscle cells in the vessel can be recreated (**d**)

Since the medial layer has a distinct three-dimensional architecture that changes from cell layer to cell layer, a different method of building up the multilayer tissue needs to be used from that described previously. Each cell layer has its own arrangement in relation to the others. Within each layer, however, the cells are organized in parallel rows so that they are all running in the same direction (Fig. 7). This is a common organizational motif that has been studied using microfabricated channels in substrates for cell monolayers. In order to build up the three-dimensional architecture that could lead to an engineered, small-diameter blood vessel with proper strength and functionality, these individual layers can be stacked on top of each other to build a medial layer with the proper number of layers and helical pitch between layers. The ability to control the number of layers as well as the pitch between layers is important because it is known that the pitch and thickness can change between vessels found in different tissues.

To control smooth muscle cell (SMC) organization PDMS scaffolds were created with parallel running channels. Channel widths of 20, 50 and 80 μm were investigated to determine an optimal width. The depth of the grooves was kept at 5 μm. This depth is shallow enough to influence a monolayer of cells. The effects of such a pattern on the organization of SMCs and their subcellular components, such as F-actin filaments, had never been investigated.

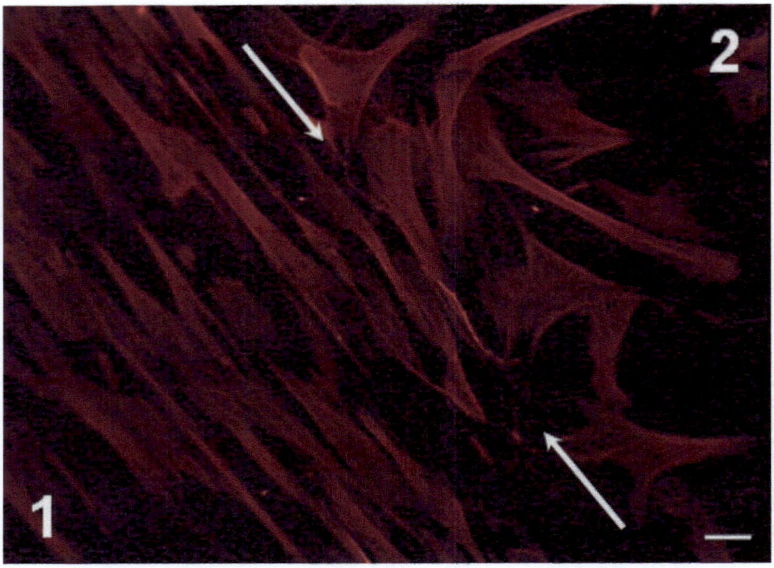


**Fig. 8** Rhodamine-phalloidin staining of the F-actin of SMCs on a PDMS substrate. Area 1 is patterned with 48-μm wide channels. Area 2 is unpatterned. The *white arrows* indicate the border between the two surfaces. The actin in the cells in area 1 have a high order of alignment while those on area 2 have a random network of actin filaments that are not organized as a population

All channel widths investigated aligned the SMCs in relation to the pattern. Quantification of the alignment showed 95% alignment on the microfabricated substrates. The aspect ratio (the ratio of the long axis of the cell to the short axis) was significantly different on each of the channel widths. Cells on the thinnest channels had an aspect ratio (~14) comparable to SMCs found in vivo. Immunofluorescence staining also showed a highly aligned F-actin network in the cells. Smooth muscle cells grown on flat PDMS substrates had no organization and had very unordered F-actin networks. This is very important because alignment of the actin cytoskeleton is required for directional contraction. By aligning the network of actin filaments throughout the population of cells, the entire unit will act to contract in a single direction (Fig. 8).

An important part of these stacked membranes will be the ability for nutrients to diffuse in through the layers and cellular waste to diffuse out so the cells in interior layers can survive. In traditional scaffolds, where the volume of the scaffold is made from one piece of material instead of being built up layer by layer, diffusion of nutrients to the center of the scaffold is accomplished through a porous network in the material. Without creating a porous network between the layers in the stacked membranes, the patterned cells in the interior of the construct will die due to inadequate nutrient supply. Creating this porous network will require creating holes in the membranes that will act as the pores. The topography of the membranes, however, is the key to organizing the cells in the appropriate manner. Disruption of the surface topography with pores could lead to a loss in cell patterning. The pores

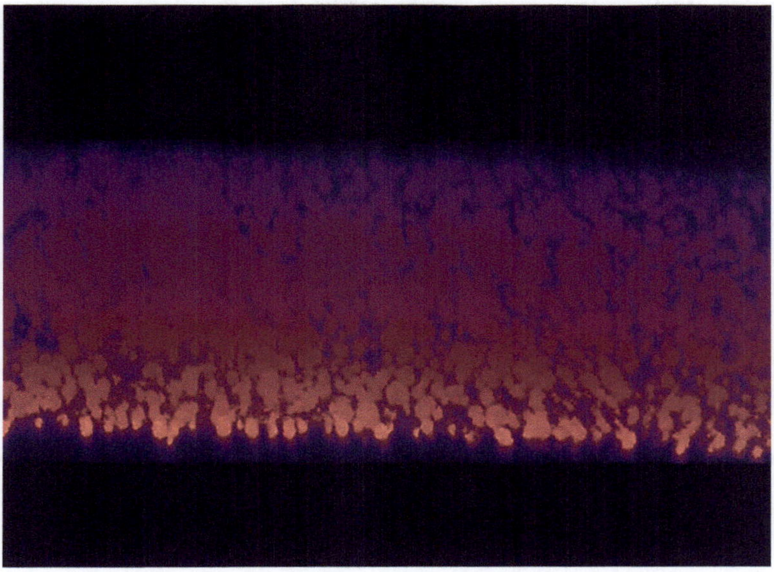

**Fig. 9** Confocal laser scanning microscopy 3D rendering of PCL-patterned scaffold containing PCL nanospheres mixed into at a ratio of 4:1 PCL:PLGA spheres. Texas Red was mixed into the PLGA during the emulsion process in order to visualize them

need to be incorporated into the membranes without disrupting the microscale topography. By creating nano-size pores, the feature size of the pores is reduced several orders of magnitude below that of the surface pattern.

To create polymer-based nanoporous scaffolds that are surface patterned at the microscale, a melt molding technique with the biopolymer poly-caprolactone (PCL) was chosen (Sarkar et al. 2006). Melt molding the PCL to a master transfers the surface pattern of channels to the PCL membrane. PCL typically has no pores when processed through melting techniques, however. Pores need to be incorporated into the PCL during this melt molding process. Common methods to introduce porous networks into polymer membranes include leaching of salts and sugars, phase separation and free-drying. These techniques create pores with sizes in the tens of micrometers which would disrupt the pattern created on the surface of the PCL membrane and inhibit cellular organization in the desired pattern. To create nanometer-size pores, polylactic-co-glycolic acid (PLGA) nanospheres were created that could be incorporated into the PCL before melt molding and then solvent leached out of the final membrane before seeding with cells. The nanospheres are created using a double emulsion-solvent evaporation technique modified from Panyam et al. (2003). This process creates spheres with an average diameter of 370 nm. The nanospheres are mixed into the PCL prior to melt molding. The higher melting temperature of PLGA in comparison to PCL leaves the nanospheres intact during the melt molding process (Fig. 9). After the PCL is pattered and cooled, the nanospheres are leached from the scaffold in a sodium hydroxide solution. The voids left by the nanospheres

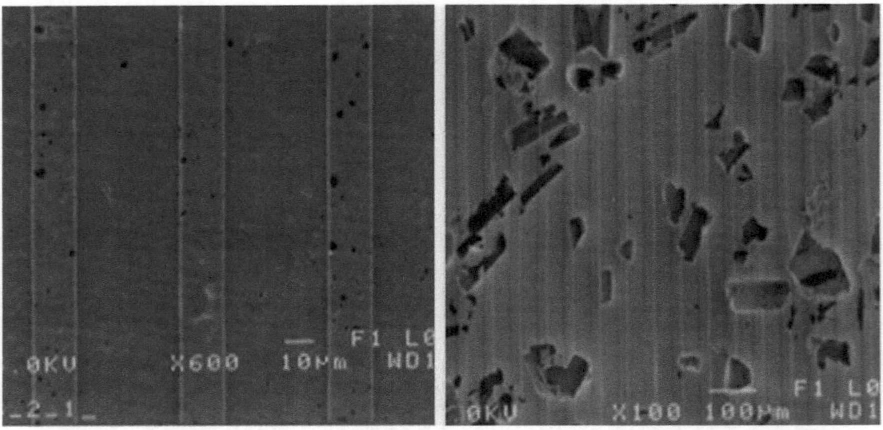


**Fig. 10** The image (SEM) on the *left* shows a membrane that has had nanopores incorporated into the PCL. The image (SEM) on the *right* shows a membrane that has had pores created through sodium bicarbonate leaching. The membrane on the *left* shows an intact pattern while the membrane on the *right* shows considerable disruption of the pattern

create a network of interconnected pores that allow diffusion through the membrane but does not disrupt the micropattern on the surface. Cells still react to the microtopography of the scaffold and align with the parallel channels. By comparison, when porous PLGA membranes were created using sodium bicarbonate crystals as the leachant, a disrupted surface pattern was generated on which the SMCs were no longer able to align (Fig. 10).

Micropatterning of scaffolds has been an important technique for tissue engineers to create controlled environments for cells in vitro in order to recreate their in vivo structure and function. With the new techniques presented here for creating three-dimensional tissue constructs, it is possible to control aspects such as porosity of the material without disrupting the physical cues with which the cells need to interact. Moving to the nanoscale to create pores now allows scientists and engineers to gain control over another size scale in the hierarchy of engineered tissues.

## 5 Nanoporous Capsules for Immunoprotection of Insulin Producing Cells

Controlling the nanoscale features of materials is not only important for creating surfaces for cells to grow on. Working at the nanometer size range means we are working at the size of individual molecules of importance to life. If we can control individual biomolecules through the intelligent engineering of materials, it may be possible to treat diseases in a manner not currently possible.

For a person suffering from Type I diabetes, the most widely used treatment is the daily injection of insulin. The use of injectable insulin has many problems for the patient including compliance, discomfort, inconvenience, and the fact that self administered insulin cannot match the natural response of insulin producing cells to food intake, exercise and stress. While whole organ (pancreas) and islet transplants are an alternative, as with any transplantation scenario, the need far outweighs the availability of supply of donor organs. As well, the patient would need to undergo a lifelong immunosuppressive regime. An alternative method under investigation is the transplantation of xenogenic or allogenic insulin producing cells sequestered in a device that could act like an artificial pancreas. This artificial organ would be made of biocompatible materials in which insulin producing cells will live and function normally. The device would be implanted into a patient in order to provide continuous insulin control through interaction with the body's biochemical signals. The device itself should be engineered to house and protect the cells from the host's immune system thereby negating the need for immunosuppressive drugs.

The success of a tissue-engineered construct when transplanted into a living being will depend foremost on the response it elicits from the host's immune system. Immune rejection of the construct is due to antibody recognition of foreign antigens present on either the cells or the scaffold material (if biological materials are used). Researchers have attempted to overcome this recognition of foreign antigens by creating immunoisolation devices to house the engineered tissue construct. These devices block antibodies from getting at the transplanted cells by creating a physical barrier to diffusion into the scaffold. The key to these immunoisolation devices is to block antibodies from entering but allow nutrients in and wastes and cellular products, most specifically insulin, out. To do this, porous materials (e.g. polymers) have been investigated as size-selective barriers. The problem with these materials, however, is that they usually exhibit a range of pore sizes that do not adequately block all antibodies. Also, these materials typically have distances of 100–200 μm over which molecules must diffuse in order to enter or leave the encapsulation device. These relatively large diffusion distances can severely hinder the dynamic response of cells to changing conditions in the body. For immunoprotection, a well-controlled pore size is needed and for dynamic control of insulin levels, as thin a barrier as possible between the host and the cells is desired. Microfabrication and nanofabrication technology has allowed for the construction of thin membranes with precisely controlled pore size and distribution.

Nanoporous capsules with 5 μm thick membranes and uniform pore dimensions and pores sizes down to 7 nm were developed for islet cell transplantation using bulk and surface micromachining techniques (Leoni and Desai 2004). Studies with these capsules have shown that cell viability and functionality were not compromised by the encapsulation (Desai et al. 2004). The porous membranes of these capsules were shown to provide sufficient insulin and glucose diffusion for nutrient exchange for the encapsulated cells. These membranes also showed an almost complete deselection of immunoglobulin G (IgG), the most abundant immunoglobulin in the body, over extended periods of time. In vivo studies done on rats with encapsulated insulinoma cells showed a reversal in diabetes and normal blood glucose levels over 2 weeks (total encapsulation time was 14 days).

Capsules with porous membranes created via the two-step anodization process used to create the bone-cell scaffolds are also being used for encapsulating insulin producing cells (La Flamme et al. 2005). This process creates pores of a slightly

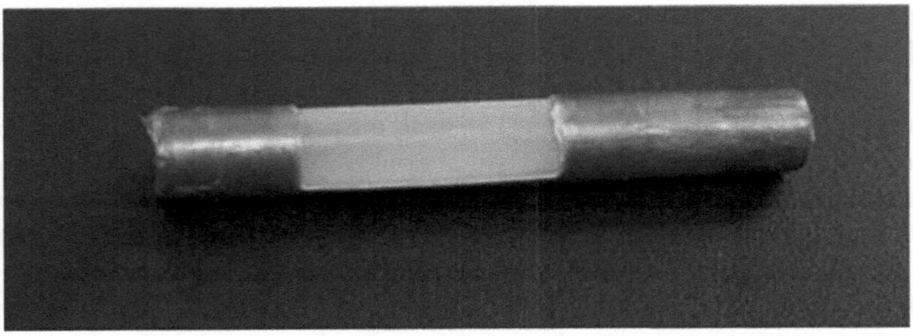

**Fig. 11** A nanoporous capsule for encapsulating insulin producing cells

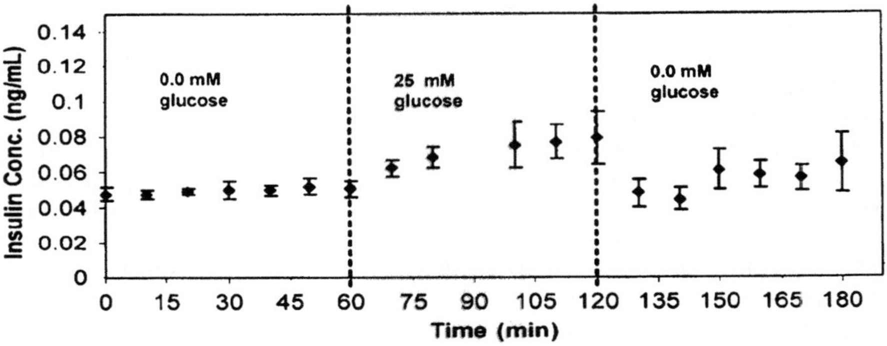


**Fig. 12** Response of a nanoporous capsule encapsulated insulionma cells to alternating low and high (*center*) glucose concentrations

larger diameter and thicker membranes than the multistep micromachining techniques described in Leoni and Desai (2004); however, they still have more precise pores and thinner membranes than polymeric devices (Fig. 11). Capsules with membranes containing 75-nm pores effectively passed glucose and hindered the passage of IgG. The diffusion coefficients of glucose and IgG through the membranes were 1.58E-06 $cm^2/s$ and 4.09E-10 $cm^2/s$, respectively. The diffusion of glucose is comparable to other encapsulation devices while the diffusion of IgG is significantly lower than with other encapsulation devices (Burczak et al. 1994; Leoni and Desai 2001). To determine the functionality of insulin producing cells in the capsule, the capsule (75 nm pores) was filled with insulinoma cells suspended in a collagen gel. The cells in the capsule were starved of glucose for approximately 24 h to bring insulin production to basal levels. The capsules were then placed in a perfusion chamber and the cells were exposed to a step-increase in glucose in the perfusion media. A resultant release of insulin from the capsules was observed indicating that the glucose enters the capsule from the outside environment and the cells can respond to the glucose by releasing insulin, which then diffuses out of the capsule into the surrounding environment. Changes in applied glucose concentrations were also seen to elicit relative changes in insulin release. High levels of glucose in the surrounding media produced higher levels of insulin release from the encapsulated cells than low levels of glucose. A step increase from low to high concentration of glucose causes increased release of insulin with a slight delay during which time it is assumed the cells are adjusting insulin production and release to the change in glucose levels (Fig. 12).

## 6 Concluding Remarks

In the future, integration of smart capabilities into these devices could turn them into complete biosensing and therapeutic platforms. Sensors to monitor cell viability, glucose and insulin levels and controlled release of agents to promote angiogenesis and biocompatibility or reduce fibrotic tissue formation around the implant would improve the ability of the implant to interact with the body. Also, the ability to transmit information outside the body so doctors can monitor the device would be attractive additions.

It is clear that the engineering of viable, active tissues and organs requires control over the environment in which the cells are going to be grown. Using techniques to image, quantify and model the environment that cells are exposed to in vivo is leading researchers to engineer that environment using a variety of micro and nanofabrication techniques. Using microtechnology, control over surface structures provides researchers with a means to systematically study how the scale and geometry of various features affected cells seeded on them. As has been shown, precise engineering of the microscale environment can create constructs that mimic native tissues. The use of nanotechnology for tissue engineering is now going through the initial stages where techniques are being refined and simplified model-systems are being used to gain insight into cellular responses to their nanoscale environment.

## References

Abrams GA, Bentley E, Nealey PF, Murphy CJ (2002) Electron microscopy of the canine corneal basement membranes. Cells Tissues Organs 170:251–257.

Abrams GA, Goodman SL, Nealey PF, Franco M, Murphy CJ (2000a) Nanoscale topography of the basement membrane underlying the corneal epithelium of the rhesus macaque. Cell Tissue Res 299:39–46.

Abrams GA, Schaus SS, Goodman SL, Nealey PF, Murphy CJ (2000b) Nanoscale topography of the corneal epithelial basement membrane and Descemet's membrane of the human. Cornea 19:57–64.

Abrams GA, Murphy CJ, Wang ZY, Nealey PF, Bjorling DE (2003) Ultrastructural basement membrane topography of the bladder epithelium. Urol Res 31:341–346.

Alberts B, Bray D, Lewis J, Raff M, Roberts K, James DW (1994) Cell junctions, cell adhesions, and the extracellular matrix. In: Molecular Biology of the Cell, pp 950–1006. New York: Garland Publishing.

Alhadlaq A, Mao JJ (2004) Mesenchymal stem cells: isolation and therapeutics. Stem Cells Dev 13:436–448.

Bissell MJ, Rizki A, Mian IS (2003) Tissue architecture: the ultimate regulator of breast epithelial function. Curr Opin Cell Biol 15:753–762.

Bissell MJ, Radisky DC, Rizki A, Weaver VM, Petersen OW (2002) The organizing principle: microenvironmental influences in the normal and malignant breast. Differentiation 70:537–546.

Boateng SY, Hartman TJ, Ahluwalia N, Vidula H, Desai TA, Russell B (2003) Inhibition of fibroblast proliferation in cardiac myocyte cultures by surface microtopography. Am J Physiol Cell Physiol 285:C171–182.

Burczak K, Fujisato T, Hatada M, Ikada Y (1994) Protein permeation through poly(vinyl alcohol) hydrogel membranes. Biomaterials 15:231–238.

Cukierman E, Pankov R, Stevens DR, Yamada KM (2001) Taking cell-matrix adhesions to the third dimension. Science 294:1708–1712.

Curtis AS, Casey B, Gallagher JO, Pasqui D, Wood MA, Wilkinson CD (2001) Substratum nanotopography and the adhesion of biological cells. Are symmetry or regularity of nanotopography important? Biophys Chem 94:275–283.

Degenaar P, Pioufle BL, Griscom L, Tixier A, Akagi Y, Morita Y, Murakami Y, Yokoyama K, Fujita H, Tamiya E (2001) A method for micrometer resolution patterning of primary culture neurons for SPM analysis. J Biochem (Tokyo) 130:367–376.

den Braber ET, de Ruijter JE, Ginsel LA, von Recum AF, Jansen JA (1998) Orientation of ECM protein deposition, fibroblast cytoskeleton, and attachment complex components on silicone microgrooved surfaces. J Biomed Mater Res 40:291–300.

Desai TA, West T, Cohen M, Boiarski T, Rampersaud A (2004) Nanoporous microsystems for islet cell replacement. Adv Drug Deliv Rev 56:1661–1673.

Deutsch J, Motlagh D, Russell B, Desai TA (2000) Fabrication of microtextured membranes for cardiac myocyte attachment and orientation. J Biomed Mater Res 53:267–275.

Dunn GA, Heath JP (1976) A new hypothesis of contact guidance in tissue cells. Exp Cell Res 101:1–14.

Fan YW, Cui FZ, Hou SP, Xu QY, Chen LN, Lee IS (2002) Culture of neural cells on silicon wafers with nano-scale surface topograph. J Neurosci Methods 120:17–23.

Fawcett DW (1986) Blood lymph vascular systems. In: Textbook of Histology, pp 367–400. Philadelphia: W.B. Saunders.

Gong D, Yadavalli V, Paulose M, Pishko M, Grimes C (2003) Controlled molecular release using nanoporous alumina capsules. Biomedical Microdevices 5:75–80.

Goodman SL, Sims PA, Albrecht RM (1996) Three-dimensional extracellular matrix textured biomaterials. Biomaterials 17:2087–2095.

Gudjonsson T, Ronnov-Jessen L, Villadsen R, Bissell MJ, Petersen OW (2003) To create the correct microenvironment: three-dimensional heterotypic collagen assays for human breast epithelial morphogenesis and neoplasia. Methods 30:247–255.

Gustafson T, Wolpert L (1999) Studies on the cellular basis of morphogenesis in the sea urchin embryo. Directed movements of primary mesenchyme cells in normal and vegetalized larvae. Exp Cell Res 253:288–295.

Hanarp P, Sutherland D, Gold J, Kasemo B (1999) Nanostructured model biomaterial surfaces prepared by colloidal lithography. Nanostructured Materials 12:429.

Hwang NS, Kim MS, Sampattavanich S, Baek JH, Zhang Z, Elisseeff J (2005) The Effects of Three Dimensional Culture and Growth Factors on the Chondrogenic Differentiation of Murine Embryonic Stem Cells. Stem Cells:113–123.

Koike N FD, Gralla O, Au P, Schechner JS, and Rakesh K Jain. (2004) Creation of long lasting blood vessels. Nature 428:138–139.

Kubosawa H, Kondo Y (1994) Quick-freeze, deep-etch studies of renal basement membranes. Microsc Res Tech 28:2–12.

La Flamme KE, Mor G, Gong D, La Tempa T, Fusaro VA, Grimes CA, Desai TA (2005) Nanoporous alumina capsules for cellular macroencapsulation: transport and biocompatibility. Diabetes Technol Ther 7:684–694.

Leoni L, Desai TA (2001) Nanoporous biocapsules for the encapsulation of insulinoma cells: biotransport and biocompatibility considerations. IEEE Trans Biomed Eng 48:1335–1341.

Leoni L, Desai TA (2004) Micromachined biocapsules for cell-based sensing and delivery. Adv Drug Deliv Rev 56:211–229.

Matsuzaka K, Walboomers XF, Yoshinari M, Inoue T, Jansen JA (2003) The attachment and growth behavior of osteoblast-like cells on microtextured surfaces. Biomaterials 24:2711–2719.

Motlagh D, Hartman TJ, Desai TA, Russell B (2003a) Microfabricated grooves recapitulate neonatal myocyte connexin43 and N-cadherin expression and localization. J Biomed Mater Res A 67:148–157.

Motlagh D, Senyo SE, Desai TA, Russell B (2003b) Microtextured substrata alter gene expression, protein localization and the shape of cardiac myocytes. Biomaterials 24:2463–2476.

Norman JJ, Desai TA (2005) Control of cellular organization in three dimensions using a microfabricated polydimethylsiloxane-collagen composite tissue scaffold. Tissue Eng 11:378–386.

Panyam J, Dali MM, Sahoo SK, Ma W, Chakravarthi SS, Amidon GL, Levy RJ, Labhasetwar V (2003) Polymer degradation and in vitro release of a model protein from poly(D,L-lactide-co-glycolide) nano- and microparticles. J Control Release 92:173–187.

Pattison MA, Wurster S, Webster TJ, Haberstroh KM (2005) Three-dimensional, nano-structured PLGA scaffolds for bladder tissue replacement applications. Biomaterials 26:2491–2500.

Pedrotty DM, Koh J, Davis BH, Taylor DA, Wolf P, Niklason LE (2005) Engineering skeletal myoblasts: roles of three-dimensional culture and electrical stimulation. Am J Physiol Heart Circ Physiol 288:H1620–1626.

Popat KC, Leary Swan EE, Mukhatyar V, Chatvanichkul KI, Mor GK, Grimes CA, Desai TA (2005) Influence of nanoporous alumina membranes on long-term osteoblast response. Biomaterials 26:4516–4522.

Powell K (2005) Stem-cell niches: it's the ecology, stupid! Nature 435:268–270.

Rahul Singhvi GSDICW (1994) Effects of substratum morphology on cell physiology. Biotechnology and Bioengineering 43:764–771.

Rhodin J (1980) Handbook of Physiology. Baltimore: Waverly Press, Inc.

Sarkar S, Lee GY, Wong JY, Desai TA (2006) Development and Characterization of a Porous Micro-Patterned Scaffold for Vascular Tissue Engineering Applications. Biomaterials 4775–4782.

Sarkar S, Dadhania M, Rourke P, Desai TA, Wong JY (2005) Vascular tissue engineering: microtextured scaffold templates to control organization of vascular smooth muscle cells and extracellular matrix. Acta Biomaterialia 1:93.

Shirato I, Tomino Y, Koide H, Sakai T (1991) Fine structure of the glomerular basement membrane of the rat kidney visualized by high-resolution scanning electron microscopy. Cell Tissue Res 266:1–10.

Sommer JR (1995) Comparative anatomy: in praise of a powerful approach to elucidate mechanisms translating cardiac excitation into purposeful contraction. J Mol Cell Cardiol 1:19–35.

Swan EE, Popat KC, Desai TA (2005a) Peptide-immobilized nanoporous alumina membranes for enhanced osteoblast adhesion. Biomaterials 26:1969–1976.

Swan EE, Popat KC, Grimes CA, Desai TA (2005b) Fabrication and evaluation of nanoporous alumina membranes for osteoblast culture. J Biomed Mater Res A 72:288–295.

Thapa A, Miller DC, Webster TJ, Haberstroh KM (2003) Nano-structured polymers enhance bladder smooth muscle cell function. Biomaterials 24:2915–2926.

van Kooten TG, Whitesides JF, von Recum A (1998) Influence of silicone (PDMS) surface texture on human skin fibroblast proliferation as determined by cell cycle analysis. J Biomed Mater Res 43:1–14.

Yamasaki Y, Makino H, Hironaka K, Hayashi Y, Shikata K, Ota Z (1990) Three-dimensional architecture of rat glomerular basement membrane by ultra-high resolution scanning electron microscopy. Acta Med Okayama 44:333–335.

# Designer Self-Assembling Peptide Nanofiber Scaffolds

**Shuguang Zhang, Hidenori Yokoi, Fabrizio Gelain, and Akihiro Horii**

## 1 Introduction

Nearly all tissue cells are embedded in 3-D microenvironment in the body surrounded by nanoscale extracellular matrix. On the other hand, nearly all tissue cells have been studied in 2-D using Petri dishes, multi-well plates or glass slides coated with various substrates. How can one reconcile the apparent disparity? Likewise, although millions of cell biology papers have been published using the 2-D culture systems, one must ask how we can be so certain the results obtained from the 2-D system truly reflect the in vivo conditions. Science, after all, is to constantly ask questions, big and small. As the late legendary Francis Crick eloquently put it "You should always ask questions, the bigger the better. If you ask big questions, you get big answers."

## 2 Two D or Not Two D

Although Petri dish has had an enormous impact on modern biology, the flat bottom nature of the Petri dish culture system, including multi-well plates, glass coverslips, etc., is less than ideal for study tissue cells for several reasons: (1) It is a 2-D system, which is in sharp contrast to the 3-D environment of natural tissues both in animals and plants. (2) The Petri dish flat surface without coating is rigid and inert, again in sharp contrast to the in vivo environment where cells intimately interact with the

S. Zhang (✉) • H. Yokoi • F. Gelain • A. Horii
Center for Biomedical Engineering, Center for Bits & Atoms
Massachusetts Institute of Technology, Cambridge, MA, USA
e-mail: shuguang@mit.edu

G.A. Silva and V. Parpura (eds.), *Nanotechnology for Biology and Medicine: At the Building Block Level*, Fundamental Biomedical Technologies,
DOI 10.1007/978-0-387-31296-5_6, 

extracellular matrix and with each other. (3) The tissue cell monolayers on coated 2-D surface, such as poly-L-lysine, collagen gels, fibronectin, laminin, and Matrigel (Kleinman et al. 1986; Kleinman and Martin 2005) as well as other synthetic materials containing segments of adhesion motifs, have only part of the cell surface attached to the materials and interact with neighboring cells. The remaining parts are often directly exposed to the culture media, unlike the majority of tissue environment where every cell intimately interacts with its neighbor cells and extracellular matrix; of course, there are exceptions to such general statement, e.g., airway epithelial cells. Thus 3-D-matrix interactions display enhanced cell biological activities. (4) The transport phenomena of 2-D and 3-D are drastically different. In 2-D culture systems, cytokines, chemokines and growth factors quickly diffuse in the media across the culture dish. This is again in sharp contrast to the in vivo environment where chemical and biological gradient diffusion systems play a vital role in signal transduction, cell–cell communications and development. (5) Cells cultured on a 2-D Petri dish are not readily transportable, that is, it is nearly impossible to move cells from one environment to another without incurring changes in the cell–material and cell–cell interactions. For example, cell collections using trypsinization or mechanically using cell scrapers may have adverse effect on cell–materials/environment interactions. In contrast, cells cultured on 3-D scaffolds are more readily transportable without significantly harming cell–material and cell–cell interactions, thus providing a significantly new way to study cell biology.

## 3 Micro- and Nanoscales: Why Are They Important?

The importance of length scales is apparent, for example, when considering the scales of trees and grasses (Fig. 1). Both are made of the same basic building blocks; sugars that are polymerized by enzymes to produce cellulose fibers. Trees, usually 20–40 cm in diameter, are common in forests. If animals are in the forest, they can either go between the trees or climb onto the trees, they cannot go cross through the trees because animals are similar in scales as the trees. On the other hand, grasses are usually 0.5 cm (commonly 0.3–1 cm in diameter) in diameter. Animals can be fully embedded in and surrounded by high grass, yet can move freely within the high grass field. Perhaps, this analogy can be extended to scaffolds of various scales for cells. Cells are commonly micrometer-scales, 5–20 μm, objects. When cells are residing on microfiber polymers, they are about same scales as the microfibers. When the cells are in the extracellular matrix, they are embedded within the nanofiber matrix, which they exceed in size by a factor of ~1,000.

In the last three decades, several biopolymers, including poly-L-lactide acid (PLLA), poly-lactic-co-glycolic acid (PLGA), PLLA–PLGA copolymers and other biomaterials including alginate, agarose, collagen gels and others, have been developed to culture cells in 3-D (Ratner et al. 1996; Lanza et al. 2000; Yannas 2001; Atala and Lanza 2002; Hoffman 2002; Palsson et al. 2003). These culture systems have significantly advanced our understanding of cell–material interactions and

**Fig. 1** The drastic difference in scales. Both trees and grass are made of cellulose, but they have different sizes. The trees shown on the *left* are 20–30 cm in diameter and the distances between the trees are in tens of meters. Animals cannot walk through the trees but between them. Some animals can climb on the trees (*left panel*). In analogy to cells which are ~5–20 μm in size, and can only attach to the microfibers. On the other hand, grass is about 0.5 cm in diameter. When animals walk in the grass field, they are fully surrounded by the grass, but can walk "through." In this case, it appears as animals are embedded in 3-D environment (*right panel*). In analogy to cellular growth in a nanofiber-based scaffold, cells are fully embedded within the scaffold that yet allows cells to migrate/move without much hindrance

fostered a new field of tissue engineering. Attempts have been made to culture cells in 3-D using synthetic polymers or copolymers. However, these synthetic (co)polymers are often processed into microfibers ~10–50 μm in diameter that are similar in size to most cells (~5–20 μm in diameter). Thus, cells attached on microfibers are more or less in a 2-D environment eventhough this is somewhat deviating from 2-D by some curvature imposed by the diameter of the microfibers. Furthermore, the micropores (~10–200 μm cross section) between the microfibers are often ~1,000–10,000-fold greater than the size of biomolecules including vitamins, amino acids, nutrients, proteins or drugs, which as a consequence can quickly diffuse away. In order to culture tissue cells in a truly 3-D microenvironment, the scaffold's fibers and pores should be significantly smaller than cells so that the cells are surrounded by the scaffold, similar to the extracellular environment and native extracellular matrices (Ayad et al. 1998; Kreis et al. 1999; Timpl et al. 1979; Kleinman et al. 1986; Lee et al. 1985; Oliver et al. 1987).

Animal-derived biomaterials (e.g., collagen gels, poly-glycosaminoglycans and Matrigel™) have been used as an alternative to synthetic scaffolds (Kubota et al. 1988; Kleinman et al. 1986; Lee et al. 1985; Oliver et al. 1987; Bissell et al. 2002; Schmeichel and Bissell 2003; Weaver et al. 1995; Zhau et al. 1997; Cukierman et al. 2001; Cukierman et al. 2002). However, while they do have an appropriate scale, they frequently contain residual growth factors, undefined constituents or non-quantified impurities. Due to lack of quality control resulting in lot to lot variations of these materials, it is thus very difficult to conduct a completely controlled study using these biomaterials. Additionally, impurities pose problems if such scaffolds would be

**Fig. 2** Architecture that mimics 3-D cellular architecture and tissue repair. The San Simeon Piccolo Dome in Venice, Italy. Each of the metal rods has a diameter of ~4 cm, 500 times smaller than the size of the dome with a diameter of ~20 m. Each rod also serves as a construction scaffold for building or repairing the dome that is truly embodied in three dimensions (*left panel*). When the repair and construction is completed, the scaffold is removed as shown (*right panel*)

considered for generating cells/tissue used in human therapy. Although researchers are well aware of its limitation, it is one of the few limited choices. Thus, it not only makes difficult to conduct a well-controlled study but also would pose problems if such scaffolds were ever used to grow tissues for human therapies.

Furthermore, after the cells are adapted to the new environment and start to make their own extracellular matrices, the artificial scaffolds that initially helped the cells should be gradually removed through absorption or biodegradation. This is in analogy similar to architectural constructions: after the entity is repaired or constructed, the scaffolds should be removed (Fig. 2).

An ideal 3-D culture system should be fabricated from a synthetic biological material with defined constituents. Thus the molecular designer self-assembling peptide nanofiber scaffolds may be a promising alternative. Figure 3 directly compares the Matrigel with the self-assembling peptide nanofiber scaffold. They have the same scales and similar porosity except Matrigel seems to have many particles/impurities contained within the Matrigel. The peptide nanofibers, however, do not exhibit such morphological impurities and display homogeneous structure (Fig. 3).

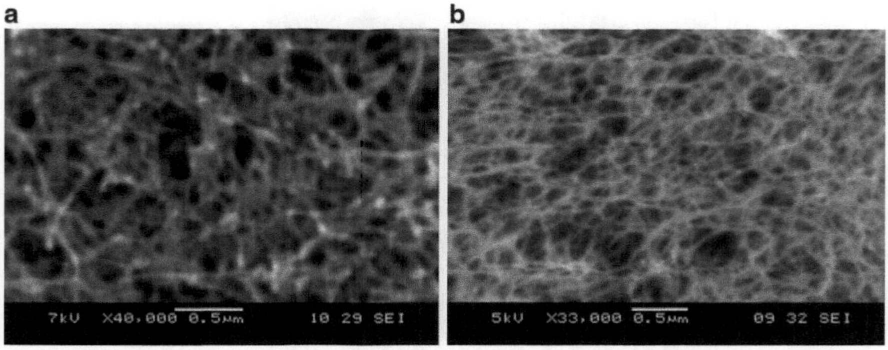


**Fig. 3** Scanning electron microscope images of Matrigel and designed self-assembling peptide nanofiber scaffold. (**a**) Matrigel™, displaying some particulate/impurities. (**b**) The self-assembling peptide RADA16-I nanofiber scaffold lacking particulate appearance and having nanopores (average 5–200 nm). Such nanopores can allow slow diffusion of small molecular drugs (1–2 nm) and proteins (2–10 nm). This is in sharp contrast to many other biopolymer microfiber materials where the micrometer pores permit drugs and proteins to diffuse rather quickly. Scale bars, 0.5 μm

# 4 Ideal Biological Materials

Although there are a number of criteria to fabricate biological scaffolds, the ideal 3-D biological scaffolds should meet several important criteria: (1) the building blocks should be derived from biological sources; (2) basic units should be amenable to design and modification to achieve specific needs; (3) exhibit a controlled rate of material biodegradation; (4) exhibit no cytotoxicity; (5) promote cell–substrate interactions, (6) afford economically scalable and reproducible material production, purification and processing; (7) be readily transportable; (8) be chemically compatible with aqueous solutions and physiological conditions; (9) elicit no or little immune responses and inflammation if used in human therapies; (10) integrate with other materials and tissue in the body.

## *4.1 Self-Assembling Peptide as Biological Material Construction Units*

In the construction industry many other parts of house, such as doors and windows can be prefabricated and then programmed assembled according to architectural plans. If we "shrink" the construction units many orders of magnitude into nanoscale, we can apply similar pre-fabrication principles to construct molecular materials and devices, through molecular programmed molecular assembly, as well as self-assembly. We limit our discussion below to three self-assembling construction units: (1) "Lego peptide" that forms well-ordered nanofiber scaffolds for 3-D cell

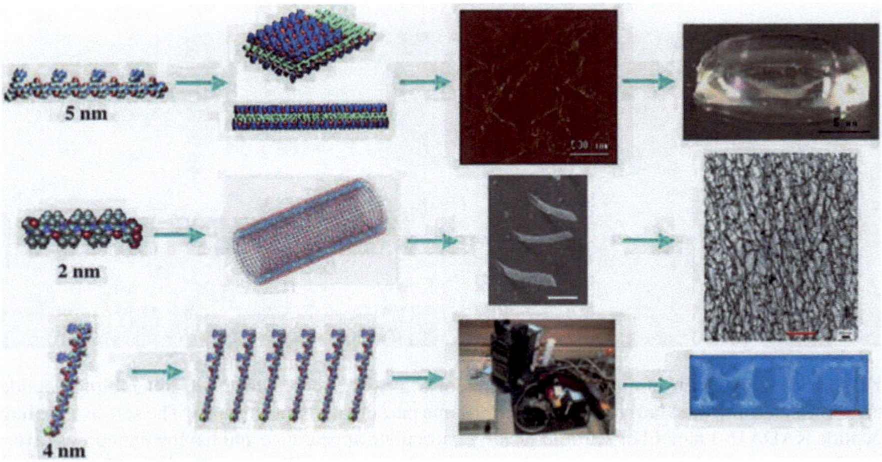


**Fig. 4** Design of various peptide materials. (*Top Panel*) Peptide Lego, also called ionic self-complementary peptide has 16 amino acids, ~5 nm in size, with an alternating polar and nonpolar pattern. They form stable β-strand and β-sheet structures, thus the side chains partition into two sides, one polar and the other nonpolar. They undergo self-assembly to form nanofibers with the nonpolar residues inside (*green*) and positively (*blue*) and negatively (*red*) charged residues form complementary ionic interactions, like a checkerboard. These nanofibers form interwoven matrices that further form a scaffold hydrogel with very high water content, >99.5% water. (*Middle Panel*) Peptide surfactant, ~2 nm in size, which has a distinct head group, either positively or negatively charged, and a hydrophobic tail consisting of six hydrophobic amino acids. They can self-assemble into nanotube and nanovesicles with a diameter of ~30–50 nm (image courtesy Steve Yang and Sylvain Vauthey). These nanotubes go on to form an interconnected network. (*Bottom Panel*) Peptide ink. This type of peptide has three distinct segments: a functional segment where it interacts with other proteins and cells; a linker segment can be flexible or stiff, and it sets the distance from the surface; and an anchor for covalent attachment to the surface. These peptides can be used as ink for an inkjet printer to directly print on a surface, instantly creating any arbitrary pattern, as shown here. Neural cells from rat hippocampal tissue grown on peptide ink substrate with defined pattern. (Images courtesy Sawyer Fuller)

culture and for regenerative medicine; (2) "lipid-like peptides" for drug, protein and gene deliveries as well as for solubilizing and stabilizing membrane proteins; (3) "peptide ink" for surface biological engineering (Fig. 4). These designed construction peptide units are structurally simple and versatile for a wide spectrum of applications as nanobiomaterials and beyond.

### 4.1.1 Discovery of Self-Assembling Peptide Scaffolds

The self-assembling peptide scaffold belongs to a class of biologically inspired materials. The first member, EAK16-II (AEAEAKAKAEAEAKAK), of the family was discovered from a segment in a yeast protein, Zuotin (Zhang et al. 1993), that was characterized as a left-handed Z-DNA binding protein (Zhang et al. 1992).

The scaffolds consist of alternating amino acids that contain 50% charged residues (Zhang et al. 1993, 1994, 1995; Holmes et al. 2000; Caplan et al. 2002; Kisiday et al. 2002; Gelain et al. 2006; Horii et al. 2007; reviewed by Gelain et al. 2007a, b). These peptides are characterized by their periodic repeats of alternating ionic hydrophilic and hydrophobic amino acids with a typical β-sheet structure. Thus, these β-sheet peptides have distinct polar and non-polar surfaces. The self-assembly event creating the peptide scaffold takes place under physiological conditions. They are like gel-sponge in aqueous solution and readily transportable to different environments. Individual fibers are ~10 nm in diameter. A number of additional self-assembling peptides including RADA16-I (AcN-RADARADARADARADA-CNH2) (shown in Fig. 3b) and RADA16-II (AcN-RARADADARARADADA-CNH2), in which arginine and aspartate residues substitute lysine and glutamate, have been designed and characterized for salt-facilitated nanofiber scaffold formation. The alanines form overlap packed hydrophobic interactions in water, a structure that is found in silk fibroin from silkworm and spiders. On the charged sides, both positive and negative charges are packed together through intermolecular ionic interactions in a checkerboard-like manner. In general, these self-assembling peptides form stable β-sheet structures in water, which are stable across a broad range of temperature, wide pH ranges in high concentration of denaturing agent urea and guanidium hydrochloride. The nanofiber density correlates with the concentration of peptide solution and the nanofiber retains extremely high hydration, >99% in water (5–10 mg/ml, w/v) (Fig. 4).

The peptide synthesis method uses conventional mature solid phase peptide synthesis chemistry. Depending on the length of the motifs, high purity of peptides can be produced at a reasonable cost. Since cost of the peptide synthesis has decreased steadily in last few years, it has become more and more affordable.

Many self-assembling peptides that form scaffolds have been reported and the numbers are still expanding (Zhang and Altman 1999; Zhang 2002). The formation of the scaffold and its mechanical properties are influenced by several factors, one of which is the level of hydrophobicity (Marini et al. 2002; Caplan et al. 2002). That is, in addition to the ionic complementary interactions, the extent of the hydrophobic amino acids, Ala, Val, Met, Ile, Leu, Tyr, Phe, Trp (or single letter code, A, V, M, I, L, Y, P, W) can significantly influence the mechanical properties of the scaffolds and the speed of their self-assembly. The higher the content of hydrophobicity, the easier it is for scaffold formation and the better for their mechanical properties (Marini et al. 2002; Caplan et al. 2002; Kisiday et al. 2002).

### 4.1.2 The Molecular Lego Peptides

Molecular-designed "Lego Peptide", at the nanometer scale, resembles the Lego bricks that have both pegs and holes in a precisely determined manner and can be programmed to assemble in well-formed structures. This class of "Lego peptide" can spontaneously assemble into well-formed nanofibers (Zhang et al. 1993). The first member of the Lego peptide was EAK16-II mentioned above and serendipitously

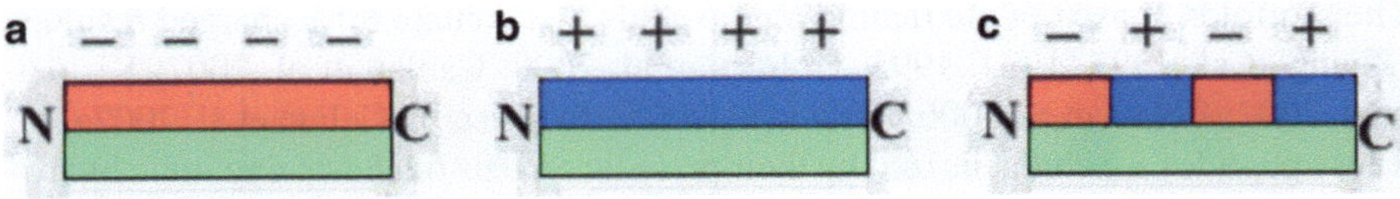


**Fig. 5** Three distinct types of self-assembling "Lego peptides." These peptides have two sides, one hydrophobic (*green*) and another hydrophilic (*red* and *blue*)

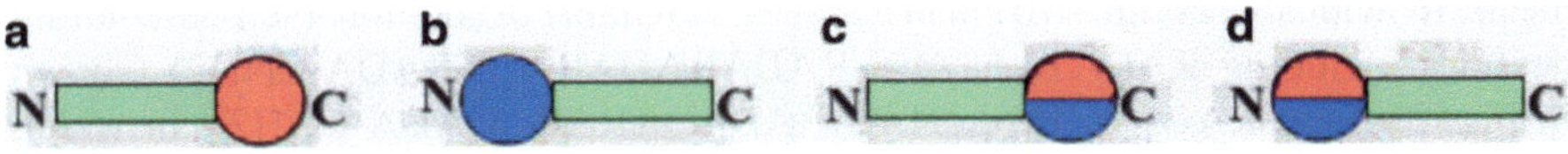


**Fig. 6** The designer lipid-like peptides. They have either negatively charged head (*red*) or positively charged head (*blue*) or mixed charged head (*half red* and *half blue* in both sites). The tails can be any hydrophobic amino acids (*green*)

discovered from Zuotin (zuo means left in Chinese, while tin is suffix to assign its peptide nature) (Zhang et al. 1992).

The Lego peptide molecules can undergo self-assembly in aqueous solutions to form well-ordered nanofibers that further associate to form nanofiber scaffolds (Fig. 4). One of them, RADA16-I, is widely used as a designed biological scaffold in contrast to other biologically derived scaffolds from animal collagen and Matrigel, which contain unspecified components in addition to known materials (Fig. 3).

Lego peptides can form stable β-strand and β-sheet structures, thus the side chains partition into two sides, one polar and the other nonpolar (Fig. 5). They undergo self-assembly to form nanofibers with the nonpolar residues inside (green) and positively (blue) and negatively (red) charged residues form complementary ionic interactions, like a checkerboard.

Since these nanofiber scaffolds contain 5–200 nm pores and have extremely high water content (>99.5% or 1–5 mg/ml), they have been used as 3-D cell-culture media. The scaffolds closely mimic the porosity and gross structure of extracellular matrices, allowing cells to reside and migrate in a 3-D environment and molecules, such as growth factors and nutrients, to diffuse in and out very slowly. These peptide scaffolds have been used for 3-D cell culture, controlled cell differentiation, tissue engineering and regenerative medicine applications.

### 4.1.3 Designer Lipid-Like Peptides

The second class of the self-assembling peptide belongs to a lipid-like molecule. These peptides have a hydrophilic head and a hydrophobic tail, much like lipids or detergents. They sequester their hydrophobic tail inside of micelle, vesicles or nanotube structures and their hydrophilic heads expose to water. At least four kinds of molecules can be made, with negative, positive, zwitterionic (±, mixed charge) heads (Vauthey et al. 2002; Santoso et al. 2002; von Maltzahn et al. 2003) (Fig. 6).

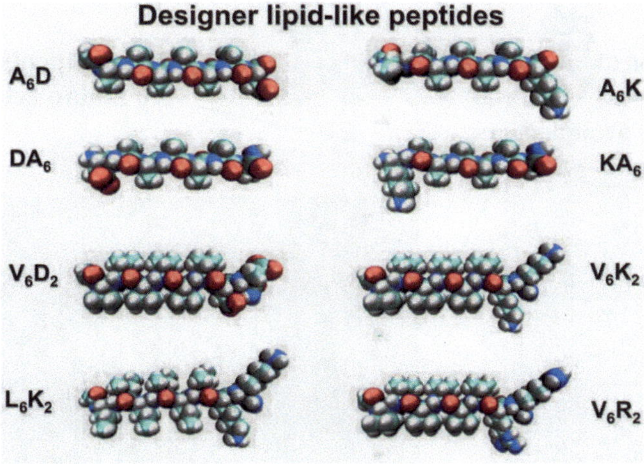


**Fig. 7** Few examples of designer lipid-like peptides. These peptides behave like lipids and surfactants that can undergo self-assembly in water to form well-ordered structures. They can also stabilize diverse membrane proteins and membrane protein complexes. The amino acids are one-letter codes, the number refers to the number of amino acid residues. Color code: *red* oxygen; *blue* nitrogen; *teal* carbon; *white* hydrogen

Several lipid-like peptides have been designed using nature's lipid as a guide. These peptides have a hydrophobic tail with various degrees of hydrophobicity and a hydrophilic head; either negatively charged aspartic and glutamic acids or positively charged lysine or histidine (Fig. 7). These peptide monomers contain 7–8 amino acid residues and have a hydrophilic head composed of aspartic acid and a tail of hydrophobic amino acids, such as alanine, valine or leucine. The length of each peptide is approximately 2 nm, similar to that of biological phospholipids (Vauthey et al. 2002; Santoso et al. 2002; von Maltzahn et al. 2003; Yang and Zhang 2006; Nagai et al. 2007; Yaghmur et al. 2007). The length can also be varied by adding more amino acids, one at a time to a desired length as shown in Fig. 8.

Although individually these lipid-like peptides have completely different composition and sequences, they share a common feature: the hydrophilic heads have 1–2 charged amino acids and the hydrophobic tails have four or more consecutive hydrophobic amino acids. For example, $A_6D$ (ac-AAAAAAD), $V_6D$ (ac-VVVVVVD) peptide has six hydrophobic alanine or valine residues from the N-terminus followed by one negatively charged aspartic acid residues, thus having two negative charges, one from the side chain and the other from the C terminus. In contrast, $V_6K_2$ (ac-VVVVVVKK) or $V_6R_2$ has six valines as the hydrophobic tail followed by two positively charged lysines or arginines as the hydrophilic head (Vauthey et al. 2002; Santoso et al. 2002; von Maltzahn et al. 2003; Yang and Zhang 2006; Nagai et al. 2007; Yaghmur et al. 2007).

Since these lipid-like peptides are not directly relevant to tissue regeneration, we will not elaborate on them further. Their applications in material sciences sprout

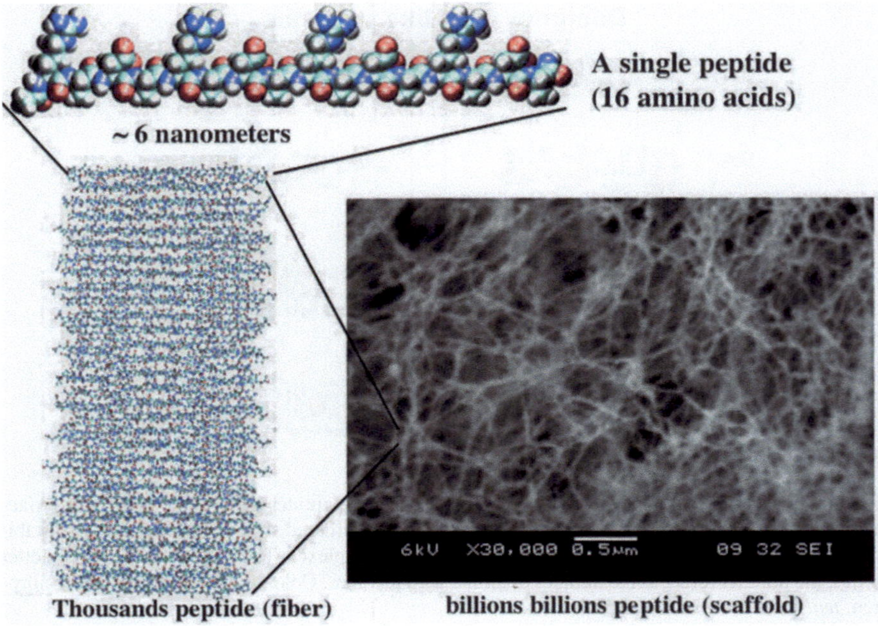


**Fig. 8** Self-assembling peptide RADA16-I nanofiber scaffold hydrogel. (*Top*) Molecular model of RADA16-I. (*Left*) Molecular model of a single RADA16-I nanofiber with its dimensions being ~6 nm long, 1.3 nm wide and 0.8 nm thick. Many individual peptides self-assemble into a nanofiber. (*Right*) SEM images of RADA16-I nanofiber scaffold. Scale bar, 0.5 μm

from their ability to stabilize a number of membrane proteins and membrane protein complexes (Kiley et al. 2005; Yeh et al. 2005; Zhao et al. 2006). Interested readers can consult the original reports and a recent summary (Zhao and Zhang 2006).

## 4.2 *Self-Assembling Peptide Nanofiber Scaffolds*

A single molecule of the ionic self-complementary peptide RADA16-I is shown in Fig. 8. Millions of peptide molecules spontaneously undergo self-assembly into individual nanofibers that further form the scaffold (Fig. 8). Between the nanofibers, there are numerous nanopores. The nanopores range from a few to a few hundred nanometers. Such structuring is similarly sized as most biomolecules and allows only slow diffusion of molecules within the scaffold and can be used to establish a molecular gradient. Figure 9 shows the individual nanofibers ranging from a few hundred nanometers to a few microns. Peptide samples in aqueous solution using an atomic force microscopy (AFM) examination display nanofibers (Fig. 9b–d), which at high resolution appear to have distinct layers in some of their segments (Fig. 9d)

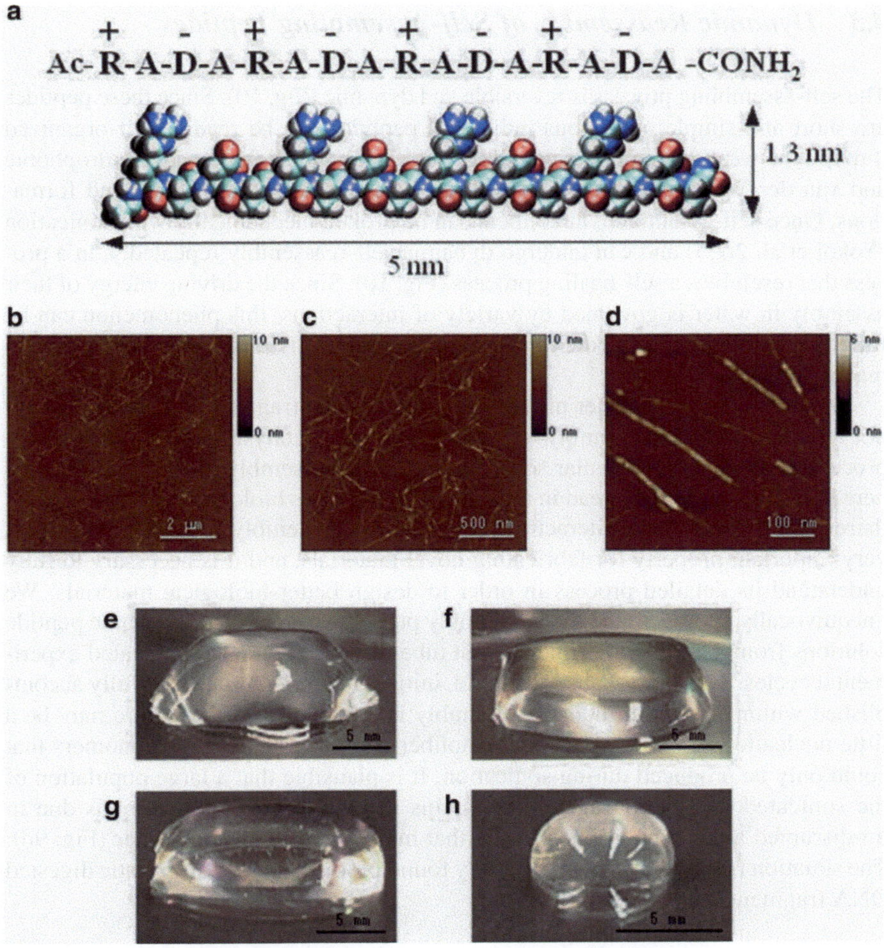


**Fig. 9** Peptide RADA16-I. (**a**) Amino acid sequence and molecular model of RADA16-I (~5 nm long, 1.3 nm wide and 0.8 nm thick). (**b–d**) AFM images of RADA16-I nanofiber scaffold. Image size, (**b**) 8×8 µm, (**c**) 2×2 µm, and (**d**) 0.5×0.5 µm. Note variability in height by ~1.3 nm of different segments within individual nanofiber suggesting its double-layered structure (**d**). (**e–h**) Photographs of RADA16-I hydrogel at various condition, (**e**) 0.5 wt.% (pH 7.5), (**f**) 0.1 wt.% (pH 7.5, Tris.HCl), (**g**) 0.1 wt.% (pH 7.5, PBS) before sonication, (**h**) reassembled RADA16-I hydrogel after four times of sonication, respectively

as distinct increase in height of about 1.3–1.5 nm can be seen; this height corresponds to the addition of a single thickness of a peptide. Figure 9e–h shows the peptide scaffold hydrogel formation at various concentrations [0.6–3 mM, or 1–5 mg/ml (w/v)] and water content (99.5–99.9%) (Yokoi et al. 2005). The scaffold hydrogel is completely transparent, which is a very important requirement for accurate image collections for uses in 3-D tissue cell cultures.

## *4.3 Dynamic Reassembly of Self-Assembling Peptides*

The self-assembling process is reversible and dynamic (Fig. 10). Since these peptides are short and simple, numerous individual peptides can be readily self-organized through the weak interactions including hydrogen bonds, ionic bonds, hydrophobic and van der Waals interactions as well as water-mediated hydrogen bond formations. Once self-assembled, nanofibers can be broken mechanically with sonication (Yokoi et al. 2005) and can undergo dynamic self-reassembly repeatedly, in a process that resembles a self-healing process (Fig. 10). Since the driving energy of their assembly in water is governed by variety of interactions, this phenomenon can be further exploited for production and fabrication of many variants of such materials.

Unlike processed polymer microfibers in which the fragments of polymers cannot readily undergo reassembly without addition of catalysts or through material processing, the supramolecular self-assembly and reassembly event we uncovered here is likely to be wide spread in many unrelated fibrous biological materials where there are numerous weak interactions involved. Self-assembly and reassembly are a very important property for fabricating novel materials, and it is necessary to fully understand its detailed process in order to design better biological materials. We unequivocally demonstrated the reassembly process since we used the same peptide solutions from a single experimental test tube throughout the four repeated experimental cycles. This remarkable and rapid, initiated within minutes and fully accomplished within ~1–2 with hours, reassembly is interesting because there may be a little nucleation for regrowth of the nanofiber from the addition of monomers that could only be produced during sonication. It is plausible that a large population of the sonicated nanofiber fragments contains many overlap cohesive ends due to un-disrupted alanine hydrophobic side that may quickly find each other (Fig. 9d). The situation is analogous and commonly found in sonicated and enzymatic digested DNA fragments.

### 4.3.1 Kinetics of Nanofiber Reassembly and a Plausible Reassembly Process

The reassembly kinetics is a function of time. Perhaps, similar to DNA reassembly, the reassembly of peptides largely depends on the concentrations of the short complementary fragments. In this case, the fragments are the sonicated peptide nanofibers with possible presence of sonicated monomers.

In order to understand the dynamic reassembly, we proposed a plausible sliding diffusion molecular model to interpret these observations of reassembly of the self-assembling RADA16-I peptides (Fig. 11). Unlike the left-handed helical structures observed in KFE8 (Marini et al. 2002), a different self-assembling peptide, no helical structures were observed for RADA16-I using AFM and transmission electron microscopy (TEM) (Holmes et al. 2000; Gelain et al. 2006).

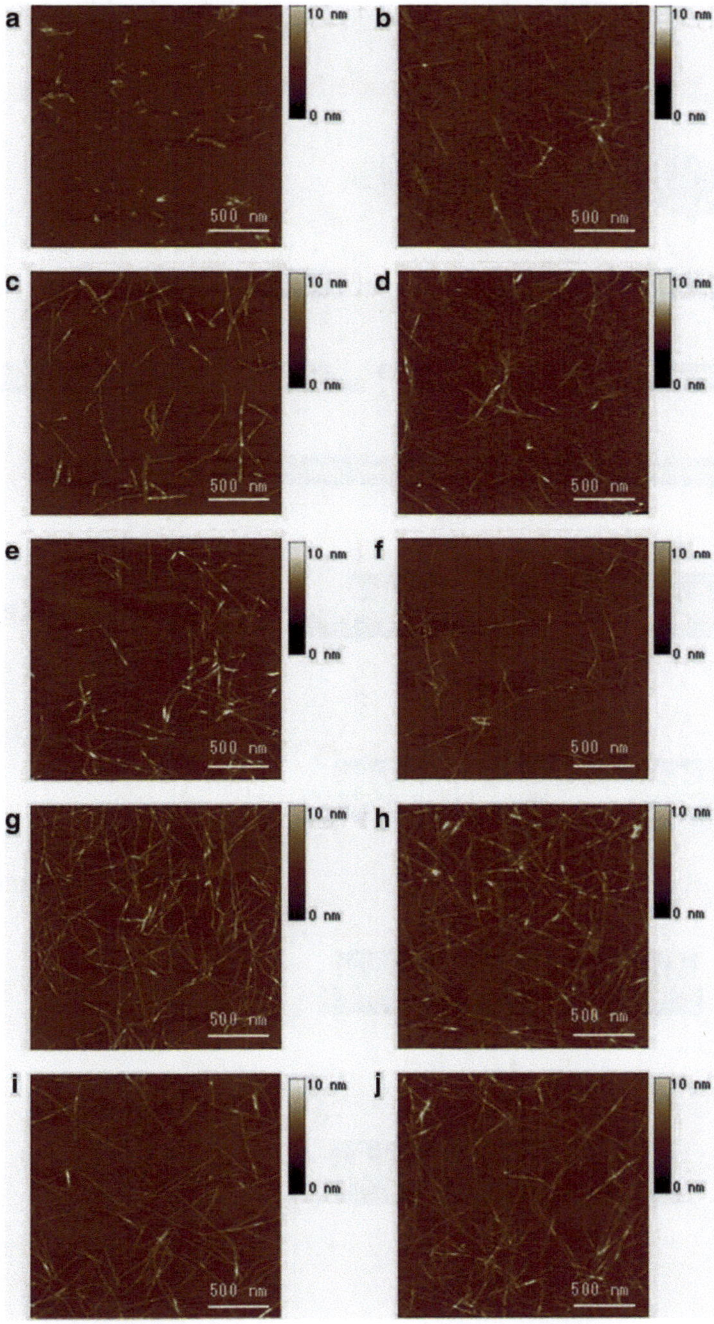


**Fig. 10** AFM images of RADA16-I nanofibers at various time points after sonication. The observations were made using AFM immediately after sample preparation. (**a**) 1 min; (**b**) 2 min; (**c**) 4 min; (**d**) 8 min; (**e**) 16 min; (**f**) 32 min; (**g**) 64 min; (**h**) 2 h; (**i**) 4 h; (**j**) 24 h after sonication. Note the elongation and reassembly of the peptide nanofibers over time. By ~1–2 h, these self-assembling peptide nanofibers have nearly fully reassembled

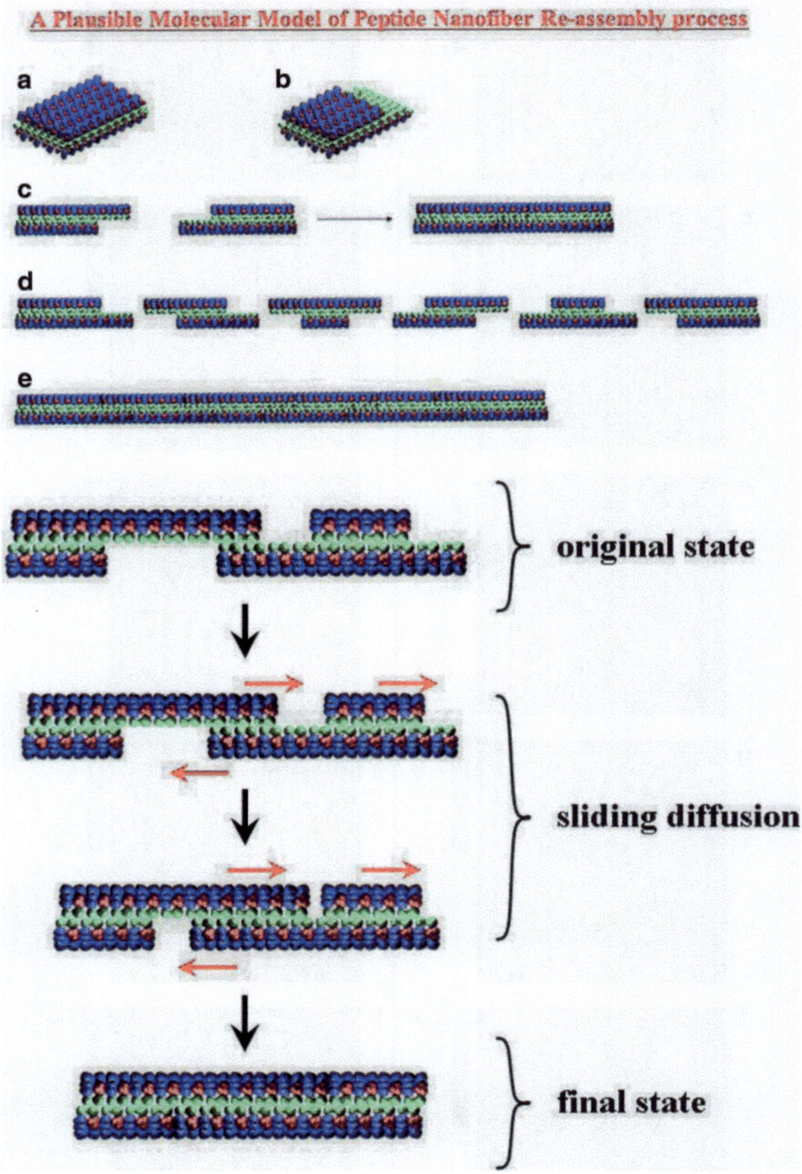
A Plausible Molecular Model of Peptide Nanofiber Re-assembly process
a
b
c
d
e
original state
sliding diffusion
final state

RADA16-I peptide can form stable β-sheet structure in water (Fig. 11). Thus, they not only form the intermolecular hydrogen bonding on the peptide backbones but they also have two distinctive sides, one hydrophobic with array of overlapping alanines (Fig. 11, green color sandwiched inside), similar as found in silk fibroin or spider silk assemblies (Pauling 1960). The other side of the backbones has negatively charged (–) aspartic acids represented as red and positively charged (+) arginines represented as blue. The alanines form packed hydrophobic interactions in water; during sonication, the hydrophobic interaction could be disrupted mechanically. However, these hydrophobic cohesive ends could find each other quickly in water since the exposure of hydrophobic alanine arrays to water is energetically unfavorable. Since the hydrophobic alanines interaction is non-specific, they can slide diffuse along the nanofiber. The same sliding diffusion phenomenon was also observed in nucleic acids where polyA and polyU form complementary base pairings that can slide diffuse along the chains (Rich and Davies 1956; Felsenfeld et al. 1957). If, however, the bases are heterogonous, containing G, A, T, C, the bases cannot undergo sliding diffusion. Likewise, if the hydrophobic side of the peptides does not always contain alanine, such as valine and isoleucine, it would become more difficult for sliding diffusion to occur due to structure constraint.

On the charged side, both positive and negative charges are packed together through intermolecular ionic interactions in a checkerboard manner (looking from the top). Likewise, the collectively complementary positive and negative ionic interactions may also facilitate the reassembly. Similar to restriction-digested DNA fragments, these nanofiber fragments could form various assemblies: blunt, semi-protruding and protruding ends. The fragments with semi-protruding and various protruding ends as well as blunt ends can reassemble readily through hydrophobic and ionic interactions.

**Fig. 11** A proposed molecular sliding diffusion model for dynamic reassembly of self-assembling RADA16-I peptides. When the peptides form stable β-sheets in water, they form intermolecular hydrogen bonds along the peptide backbones. The β-sheets have two distinctive sides, one hydrophobic with an array of alanines and the other with negatively charged aspartic acids and positively charged arginines. These peptides form anti-parallel β-sheet structures. The alanines form overlap packed hydrophobic interactions in water. On the charged sides, both positive and negative charges are packed together through intermolecular ionic interactions in a checkerboard-like manner. These nanofiber fragments can form various assemblies similar to restriction-digested DNA fragments: (**a**) blunt ends; (**b**) semi-protruding ends; (**c**) These fragments with protruding ends could reassemble readily through hydrophobic interactions; (**d**) The fragments with semi-protruding and various protruding ends; (**e**) These fragments can reassemble readily. (*Bottom*) A proposed molecular sliding diffusion model for dynamic reassembly of self-assembling a single peptide nanofiber consisting thousands of individual peptides. When the fragments of nanofiber first meet, the hydrophobic sides may not fit perfectly but with gaps (original state). However, the non-specific hydrophobic interactions permit the nanofiber to slide diffusion (sliding diffusion) along the fiber in either direction that minimizes the exposure of hydrophobic alanines and eventually fill the gaps (final state). For clarity, these β-sheets are not presented as twisted strands. Color code: *green* alanines; *red* negatively charged aspartic acids; *blue* positively charged arginines

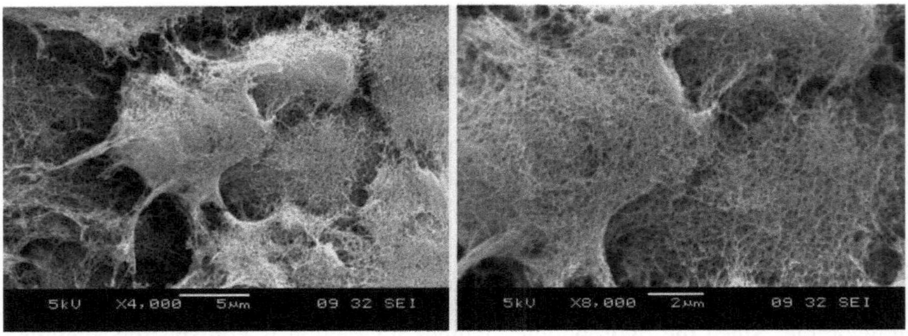


**Fig. 12** Cluster of cells are fully embedded in the self-assembling peptide nanofiber scaffold. The scales of the nanofibers are on the similar scale as the native extracellular matrices. Such 3-D cell clusters are nearly impossible to form in the 2-D culture systems

## *4.4 Self-Assembling Peptides Nanofiber Scaffold 3-D Cell Culture*

The importance of nanoscale becomes obvious in 3-D cell culture. It is clearly visible in the scanning electron microcopy (SEM) images that the cells embed in the self-assembling peptide nanofiber biological scaffolds in the truly 3-D culture (Fig. 12). Here, the cells and cell clusters intimately interact with the extracellular matrix that the cells make on their own. Since the scaffolds are made mostly of water, ~99% water at 1% peptide solid, cells can migrate freely without hindrance.

These new self-assembling peptide nanofiber biological scaffolds have become increasingly important not only in studying spatial behaviors of cells, but also in developing approaches for a wide range of innovative medical technologies including regenerative medicine (Fig. 13). One example is the use of the peptide scaffolds to support neurite growth and differentiation, neural stem cell differentiation, cardiac myocytes, bone and cartilage cell cultures. The peptide scaffolds from RADA16-I and RADA16-II form nanofiber scaffold in physiological solutions that stimulated extensive rat neurite outgrowth, and active synapses formation on the peptide scaffold was successfully achieved (Holmes et al. 2000).

## *4.5 Designer Peptides Scaffold 3-D Cell Cultures*

A variety of mammalian cells have been cultured on designer self-assembling peptide nanofiber scaffolds (Table 1). In a recent work we directly and systematically compared neural stem cell adhesion and differentiation on self-assembling RADA16-I scaffolds with other natural-based substrates including laminin, Collagen I, fibronectin and some of the most commonly used synthetic biomaterials in tissue

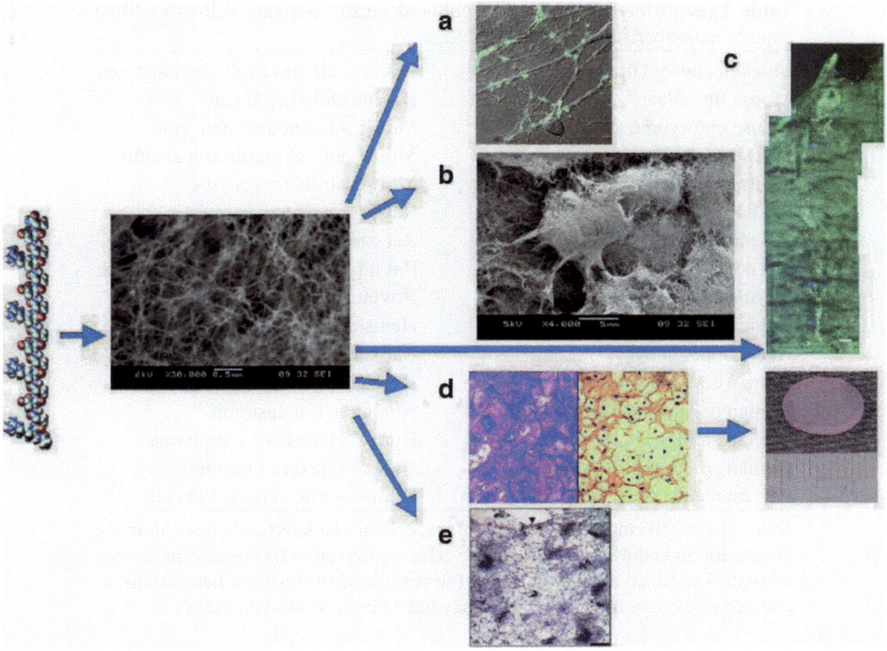


**Fig. 13** From designer self-assembling peptides to nanofibers to scaffolds to tissues. (**a**) Active synapses on the peptide surface. Primary rat hippocampal neurons form active synapses on peptide scaffolds. The confocal images shown in bright discrete *green dot* labeling are indicative of synaptically active sites after incubation of neurons with the fluorescent lipophilic probe FM1-43. The active synapses on the peptide scaffold are fully functional, indicating that the peptide scaffold is a permissible material for neurite outgrowth and active synapse formation. (**b**) Adult mouse neural stem cells embedded in 3-D scaffold. (**c**) Brain damage repair in hamster. The peptide scaffold was injected into the optical nerve area of brain that was first severed with a knife. The cut was sealed by the migrating cells after 2 days. A great number of neurons form synapses (image courtesy of Rutledge Ellis-Behnke). (**d**) Chondrocytes in the peptide KLD12 (KLDLKLDLKLDL) scaffold and cartilage formation. The trypan *blue-stained* chondrocytes show abundant glycosaminoglycan production (*left panel*), while antibody to type II collagen demonstrates abundant Type II collagen production (*right panel*). A piece of pre-molded cartilage with encapsulated chondrocytes in the peptide nanofiber scaffold. The cartilage formed over a 3–4 week period after the initial seeding of the chondrocytes (image courtesy of John Kisiday). (**e**) Von Kossa staining showing transverse sections of primary osteoblast cells on HA-PHP-RADA16-I self-assembling peptide nanofiber scaffold. Scale bar=0.1 mm. The intensely stained *black areas* represent bone nodules forming (image courtesy of Maria Bokhari et al. 2005)

engineering, such as poly- (DL-lactic acid), poly- (lactide-co-glycolide acid) and poly- (capro-lactone acid) (Gelain et al. 2007a, b). While natural-derived substrates showed the best performances, RADA16-I scaffold coaxed neural stem cell differentiation and survival to a similar degree of the other synthetic biomaterials.

Although self-assembling peptides are promising scaffolds, they show no specific cell interaction because their sequences are not naturally found in living

**Table 1** A variety of tissue cells cultured on the designer self-assembling peptide nanofiber scaffolds

| | |
|---|---|
| Chicken embryo fibroblast | Bovine calf and adult chondrocytes |
| Mouse fibroblast | Bovine endothelial cells |
| Mouse embryonic stem cells | Mouse adult neural stem cells |
| Mouse cerebellum granule cells | Mouse and rat hippocampal cells |
| Mouse mesenchymal stem cells | Mouse cardiac myocytes |
| Rat adult liver progenitor cells | Rat liver hepatocytes |
| Rat pheochromocytoma | Rat cardiac myocytes |
| Rat neural stem cells | Rat hippocampal neural tissue slice |
| Bovine osteoblasts | Bovine endothelium cells |
| Chinese hamster ovary | Hamster pancreas cells |
| Horse bone marrow | Rat keratinocytes |
| Human cervical carcinoma | Human osteosarcoma |
| Human hepato-cellular carcinoma | Human neuroblastoma |
| Human embryonic kidney | Human Hodgkin's lymphoma |
| Human epidermal keratinocytes | Human foreskin fibroblast |
| Human neural stem cells | Human aortic endothelial cells |

*Note*: These cells include stable cell lines, primary isolated cells from animals, progenitor and adult stem cells. These cells are known to be cultured in various laboratories. Since this compilation, the peptide scaffolds have been commercialized and many more cell types may have been cultured on them

systems. The next step is to directly couple biologically active and functional peptide motifs to generate the second generation of designer scaffolds that would significantly improve their interactions with cells and tissues.

The simplest way to incorporate the functional motifs is to directly synthesize them by extending the motifs on to the self-assembling peptides themselves (Fig. 14). The functional motifs are on the C-termini, since peptide synthesis starts from C-termini to avoid deletion during synthesis. Usually a spacer comprising two glycine residues is added to guarantee a flexible and correct exposure of the motifs to cell surface receptors. Different functional motifs in various ratios can be incorporated in the same scaffold. Upon exposure to solution with neutral pH, the functionalized sequences self-assemble leaving the added motifs flagging on both sides of each nanofiber (Fig. 14). These nanofibers with functional motifs take part in the overall scaffold thus giving microenvironments functionalized with specific biological stimuli (Fig. 14).

The self-assembling peptide scaffolds with functional motifs can be commercially produced with a reasonable cost. Thus, this method can be readily adopted for widespread uses including study how cell interact with their local- and microenvironments, cell migrations in 3-D, tumor and cancer cells' interactions with normal cells, cell process and neurite extensions, cell-based drug test assays and other diverse applications.

We have produced different designer peptides from a variety of functional motifs with different lengths (Gelain et al. 2006; Horii et al. 2007). We showed that the addition of motifs to the self-assembling peptide RADA16-I did not inhibit

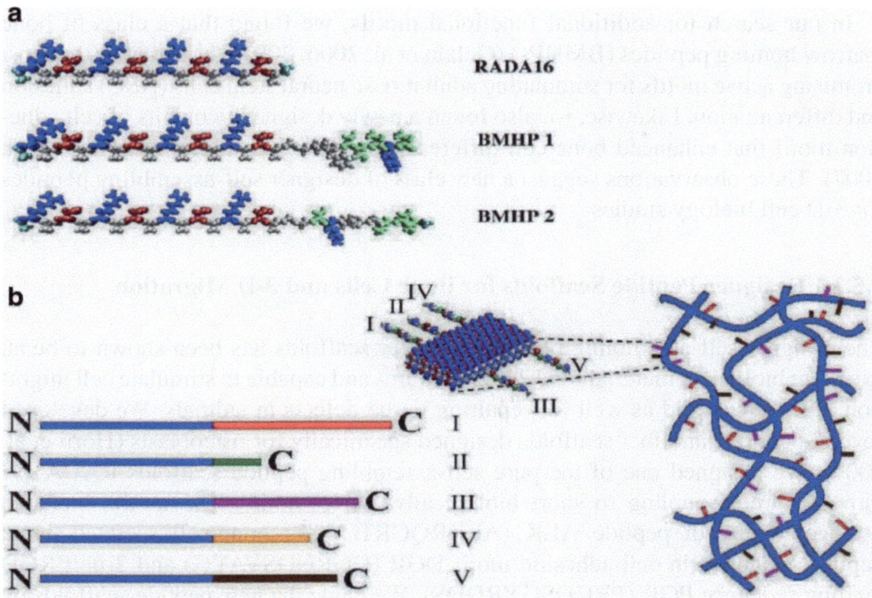


**Fig. 14** Molecular and schematic models of the designer peptides and of the scaffolds. (**a**) Molecular models of RADA16, RADA16-Bone Marrow Homing Peptide 1 (BMHP1) and RADA16-Bone Marrow Homing Peptide 2 (BMHP2). RADA16 is an alternating16-residue peptide with basic arginine (*blue*), hydrophobic alanine (*white*) and aspartic acid (*red*). These peptides self-assemble once exposed to physiological pH solutions or salt. The alanines of the RADA16 providing hydrophobic interaction are on one side of the peptide, and the arginines and aspartates form complementary ionic bonds on the other. The BMHP1 and BMHP2 motifs were directly extended from RADA16 with two glycine spacers and are composed of a lysine (*blue*), serine and threonine (*green*) and different hydrophobic (*white*) residues. Neutral polar residues are drawn in *green*. (**b**) Schematic models of several different functional motifs (different *colored bars*) could be extended from RADA16 (*blue bars*) in order to design different peptides (I, II, III, IV and V). They can be combined in different ratios. A schematic model of a self-assembling nanofiber scaffold with combinatorial motifs carrying different biological functions is shown

self-assembling properties and nanofiber formations through mixing the modified peptides with the original RADA16-I. Although their nanofiber structures appear to be indistinguishable from the RADA16-I scaffold, the appended functional motifs significantly influenced cell behaviors.

Using the designer self-assembling peptide nanofiber system, every ingredient of the scaffold can be defined and combined with various functionalities including the soluble factors. This is in sharp contrast with 2-D systems where cells attach and spread only on the planar surface; cells residing in a 3-D environment can interact with their extracellular matrix receptors to functional ligands appended to the peptide scaffolds. It is possible that higher tissue architectures with multiple cell types, rather than monolayers, could be constructed using these designer 3-D self-assembling peptide nanofiber scaffolds.

In our search for additional functional motifs, we found that a class of bone marrow homing peptides (BMHPs) (Gelain et al. 2006, 2007a, b) is one of the most promising active motifs for stimulating adult mouse neural stem cells (NSC) adhesion and differentiation. Likewise, we also found a newly designed two units of cell adhesion motif that enhanced bone cell differentiation and 3-D migration (Horii et al. 2007). These observations suggest a new class of designer self-assembling peptides for 3-D cell biology studies.

### 4.5.1 Designer Peptide Scaffolds for Bone Cells and 3-D Migration

The designer self-assembling peptide nanofiber scaffolds has been shown to be an excellent biological material for 3-D cell cultures and capable to stimulate cell migration into the scaffold as well for repairing tissue defects in animals. We developed several peptide nanofiber scaffolds designed specifically for osteoblasts (Horii et al. 2007). We designed one of the pure self-assembling peptide scaffolds RADA16-I through direct coupling to short biologically active motifs. The motifs included osteogenic growth peptide ALK (ALKRQGRTLYGF) bone-cell secreted-signal peptide, osteopontin cell adhesion motif DGR (DGRGDSVAYG) and 2-unit RGD binding sequence PGR (PRGDSGYRGDS). We made the new peptide scaffolds by mixing the pure RADA16-I and designer peptide solutions, and we examined the molecular integration of the mixed nanofiber scaffolds using AFM. Compared to pure RADA16-I scaffold, we found that these designer peptide scaffolds significantly promoted mouse pre-osteoblast MC3T3-E1 cell proliferation. Moreover, alkaline phosphatase (ALP) activity and osteocalcin secretion, which are early and late markers for osteoblastic differentiation, were also significantly increased. We demonstrated that the designer, self-assembling peptide scaffolds promoted the proliferation and osteogenic differentiation of MC3T3-E1. Under the identical culture medium condition, confocal images unequivocally demonstrated that the designer PRG peptide scaffold stimulated cell migration into the 3-D scaffold (Fig. 15) (Horii et al. 2007). Without the modified motif, cells did not migrate into 3-D.

## *4.6 Why Designer Self-Assembling Peptide Scaffolds?*

One may ask why one should choose designer self-assembling peptide scaffolds while there are a large number of biomaterials on the market and some have already been approved by the U.S Food and Drug Administration (FDA). The advantage of using the designer peptide nanofiber scaffolds is severalfold. (1) One can readily modify the designer peptides at the single amino acid level at will, inexpensively and quickly. This level of modification is impossible with Matrigel and other polymer scaffolds. (2) Unlike Matrigel, which contains unknown ingredients and quality that varies from batch to batch, the designer self-assembling peptide scaffolds belong to a class of synthetic biological scaffolds that contains pure components and every ingredient is completely defined. (3) Because these designer peptide scaffolds are

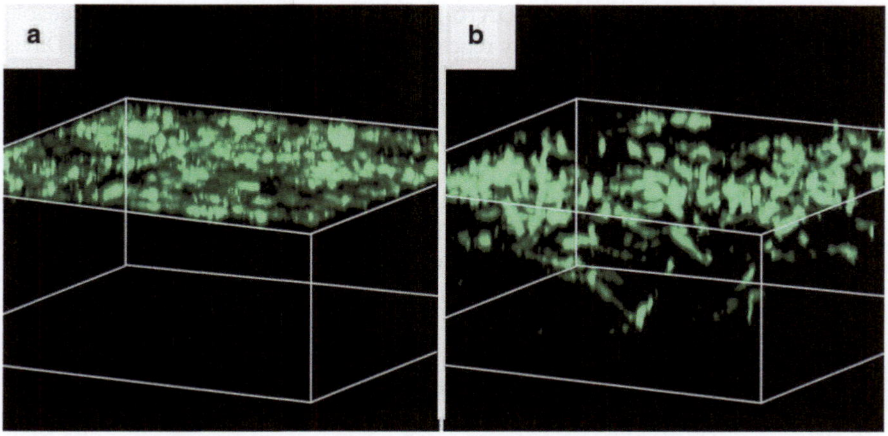


**Fig. 15** Reconstructed image of 3-D confocal microscope image of culturing PRG scaffolds. The vertical depth is ~400 μm. (**a**) 10% PRG scaffold where the cells stayed on the surface of the scaffold. (**b**) 70% PRG scaffold where the cells migrated into the scaffold. There is a drastic cell migration into the scaffold with higher concentration of the PRG motif

well defined using the known motifs, it can be used to study controlled gene expression or cell signaling process. Thus these new designer nanofiber scaffolds proved to be promising tools to study cell signal pathways in a selective way not possible with any substrates including Matrigel and collagen gels, which result in confusing cell signaling activation. (4) The initiation of the self-assembly process is through change of ionic strength at the physiological conditions without temperature influence. This is again unlike collagen gels, for which the gelation is through change of temperature that can sometimes induce unknown biological process including cold or heat shocks. (5) These scaffolds provide the opportunity to incorporate a number of different functional motifs and their combinations to study cell behavior in a well-defined ECM-analog microenvironment, not only without any chemical cross-link reactions but also fully bio-reabsorbable scaffolds. It should be noted that various animals, including mice, rats, hamsters, rabbits, goats, monkey, pigs and horses, have been exposed to peptide nanofiber scaffolds. The tests conducted in different academic laboratories, commercial laboratories, as well as biomaterials and medical device companies around the world indicate that peptide nanofiber scaffolds appeared to be harmless to animal health.

## 5 Beyond 3-D Cell Cultures

Researchers in neuroscience have a strong desire to study neural cell behaviors in 3-D and to fully understand their connections and information transmission (Edelman and Keefer 2005). Since the building blocks of designer peptide scaffolds are natural L-amino acids, the RADA16-I has been shown not to elicit noticeable immune response, nor inflammatory reactions in animals (Zhang et al. 2005; Davis et al. 2005; Ellis-Behnke et al. 2006 a, b), the degraded products can be reused by the body, they may also be useful as a bio-reabsorbable scaffold for neural repair and neuroengineering to alleviate and to treat a number of neuro-trauma and neuro-degeneration diseases.

In a recent work led by Richard Lee, mouse embryonic stem cells were suspended in RADA16-II peptide scaffold solutions and injected in the myocardium of 10-weeks-old mice (Davis et al. 2005). In that study it has been demonstrated that self-assembling peptides can be injected into the myocardium to create 3-D microenvironment. After 7, 14 and 28 days these microenvironments recruit both endogenous endothelial and smooth muscle cells, and exogenously injected cells survive in the microenvironments: self-assembling peptides can thus create injectable microenvironments that promote vascularization.

In addition Lee's group also developed an appealing drug delivery strategy by using a biotinylated version of RADA-II to demonstrate a slow release of Insulin-like growth factor 1 (IGF-1) in infarctuated rat myocardia (Davis et al. 2006). The biotin sandwich strategy allowed binding of IGF-1 and did not prevent self-assembly of the peptides into nanofibers within the myocardium. In conjunction with cardiomyocytes transplantation the strategy showed that cell therapy with IGF-1 delivery by biotinylated nanofibers significantly improved systolic function after experimental myocardial infarction.

Ellis-Behnke and colleagues showed that self-assembling peptide material is a promising scaffold for neural regeneration medicine (Ellis-Behnke et al. 2006 a, b). In vivo application to brain wounds was carried out using postnatal day-2 Syrian hamster pups. The optic tract within the superior colliculus (SC) was completely severed with a deep knife wound, extending at least 1 mm below the surface. At surgery, ten animals were treated by injection into the wound with 10–30 μl of 1% RADA16-I in 99% water (w/v). Control animals with the same brain lesion included 3 with isotonic saline injection (10 μl), and numerous additional cases, including 10 in which the dye Congo red was added into the peptide scaffold, and 27 earlier animals with knife cuts and no injection surviving 6–9 days. Animals were sacrificed at 1, 3, 6, 30 and 60 days for brain examinations. Histological specimen examinations revealed that only in the peptide scaffold-injected animals, but not in untreated animals, the brain tissue appears to have reconnected itself together in all survival times. Additionally, axons labeled from their retinal origin with a tracer molecule were found to have grown beyond the tissue bridge, reinnervating the SC caudal to the lesion. Most importantly, functional tests proved a significant restoration of visual function in all peptide scaffold-treated animals.

Ellis-Behnke and colleagues during the brain surgery experiments found that the peptide nanofiber scaffold hydrogel could also stop bleeding in less than 15 s (Ellis-Behnke et al. 2007). This is unlikely to be the conventional blood clogging mechanism because it takes place so rapidly. The molecular mechanism of speedy stopping bleeding still remains to be uncovered. It is plausible that the nanofibers self-assembled at the site quickly self-assembled into a dense mesh nanofiber network sponge that instantly blocked the rushing of the liquid. It may be perhaps nano-mechanics rather than biochemistry.

The development of new biological materials, particularly those biologically inspired nanoscale scaffolds mimicking in vivo environment, that serve as permissive substrates for cell growth, differentiation and biological function is an actively pursued area, which in turn could significantly advance regenerative medicine. These materials will be useful not only for further understanding of cell biology in 3-D environment but also for advancing medical technology, regenerative biology and medicine.

**Acknowledgments** We gratefully acknowledge the supports by grants from Olympus Corp., Japan; Menicon, Ltd, Japan and fellowship to FG from Fondazione Centro San Raffaele del Monte Tabor, Milan, Italy.

## References

Atala A, Lanza R (2002) Methods of Tissue Engineering Academic Press.

Ayad S B-HR, Humphries M, Kadler K, Shuttleworth A (1998) The Extracellular Matrix Factsbook. 2nd Ed Academic Press.

Bissell MJ (1981) The differentiated state of normal and malignant cells or how to define a "normal" cell in culture. Int Rev Cytol 70:27–100.

Bissell MJ, Radisky DC, Rizki A, Weaver VM, Petersen OW (2002) The organizing principle: microenvironmental influences in the normal and malignant breast. Differentiation 70: 537–546.

Bokhari MA, Akay G, Zhang S, Birch MA (2005) The enhancement of osteoblast growth and differentiation in vitro on a peptide hydrogel-polyHIPE polymer hybrid material. Biomaterials 26:5198–5208.

Caplan M, Schwartzfarb E, Zhang S, Kamm R, Lauffenburger D (2002) Control of self-assembling oligopeptide matrix formation through systematic variation of amino acid sequence. Biomaterials 23:219–227.

Cukierman E, Pankov R, Yamada KM (2002) Cell interactions with three-dimensional matrices. Curr Opin Cell Biol 14:633–639.

Cukierman E, Pankov R, Stevens D, Yamada K (2001) Taking cell-matrix adhesions to the third dimension. Science 294:1708–1712.

Davis ME, Motion JP, Narmoneva DA, Takahashi T, Hakuno D, Kamm RD, Zhang S, Lee RT (2005) Injectable self-assembling peptide nanofibers create intramyocardial microenvironments for endothelial cells. Circulation 111:442–450.

Davis ME, Hsieh PC, Takahashi T, Song Q, Zhang S, Kamm RD, Grodzinsky AJ, Anversa P, Lee RT (2006) Local myocardial insulin-like growth factor 1 (IGF-1) delivery with biotinylated peptide nanofibers improves cell therapy for myocardial infarction. Proc Natl Acad Sci U S A 103:8155–8160.

Edelman DB, Keefer EW (2005) A cultural renaissance: in vitro cell biology embraces three-dimensional context. Exp Neurol 192:1–6.

Ellis-Behnke RG, So K-F, Zhang S (2006) Molecular repair of the brain using self-assembling peptides. Chemistry Today 24:42–45.

Ellis-Behnke RG, Liang YX, Tay DKC, Kau PWF, Schneider GE, Zhang S, Wu W, So K.-F (2006) Nano hemostat solution: immediate hemostasis at the nanoscale. Nanomedicine. Nanotechnology, Biology & Medicine 2:207–215.

Ellis-Behnke RG, Liang YX, You SW, Tay DK, Zhang S, So KF, Schneider GE (2006a) Nano neuro knitting: peptide nanofiber scaffold for brain repair and axon regeneration with functional return of vision. Proc Natl Acad Sci U S A 103:5054–5059.

Ellis-Behnke RG, Liang YX, Tay DK, Kau PW, Schneider GE, Zhang S, Wu W, So KF (2006b) Nano hemostat solution: immediate hemostasis at the nanoscale. Nanomedicine 2:207–215.

Felsenfeld G, Davies DR, Rich A (1957) Formation of a three-stranded polynucleotide molecule. J Am Chem Soc 79:2023–2024.

Gelain F, Horii A, Zhang S (2007a) Designer self-assembling peptide scaffolds for 3-d tissue cell cultures and regenerative medicine. Macromol Biosci 7:544–551.

Gelain F, Bottai D, Vescovi A, Zhang S (2006) Designer self-assembling peptide nanofiber scaffolds for adult mouse neural stem cell 3-dimensional cultures. PLoS One 1:e119.

Gelain F, Lomander A, Vescovi AL, Zhang S (2007b) Systematic studies of a self-assembling peptide nanofiber scaffold with other scaffolds. J Nanosci Nanotechnol 7:424–434.

Hoffman AS (2002) Hydrogels for biomedical applications. Adv Drug Deliv Rev 43:3–12.

Holmes TC, de Lacalle S, Su X, Liu G, Rich A, Zhang S (2000) Extensive neurite outgrowth and active synapse formation on self-assembling peptide scaffolds. Proc Natl Acad Sci USA 97:6728–6733.

Horii A, Wang X, Gelain F, Zhang S (2007) Biological designer self-assembling peptide nanofiber scaffolds significantly enhance osteoblast proliferation, differentiation and 3-D migration. PLoS One 2:e190.

Kiley P, Zhao X, Vaughn M, Baldo M, Bruce BD, Zhang S (2005) Self-assembling peptide detergents stabilize isolated photosystem I on a dry surface for an extended time. PLoS Biol 3:1181–1186.

Kisiday J, Jin M, Kurz B, Hung H, Semino C, Zhang S, Grodzinsky AJ (2002) Self-assembling peptide hydrogel fosters chondrocyte extracellular matrix production and cell division: implications for cartilage tissue repair. Proc Natl Acad Sci U S A 99:9996–10001.

Kleinman HK, McGarvey ML, Hassell JR, Star VL, Cannon FB, Laurie GW, et al. (1986) Basement membrane complexes with biological activity. Biochemistry 25:312–318.

Kleinman HK, Martin GR (2005) Matrigel: basement membrane matrix with biological activity. Semin Cancer Biol 15:378–386.

Kreis T, Vale R (1999) Guide book to the extracellular matrix, anchor, and adhesion proteins. 2nd ed. Oxford, UK: Oxford University Press.

Kubota Y, Kleinman HK, Martin GR, Lawley TJ (1988) Role of laminin and basement membrane in the morphological differentiation of human endothelial cells into capillary–like structures. J Cell Biol 107:1589–1598.

Lanza R, Langer R, VacantiJ (2000) Principles of Tissue Engineering 2nd edition. Academic Press: San Diego, CA.

Lee EY, Lee WH, Kaetzel CS, Parry G, Bissell MJ (1985) Interaction of mouse mammary epithelial cells with collagen substrata: regulation of casein gene expression and secretion. Proc Natl Acad Sci U S A 82:1419–1423.

Marini DM, Hwang W, Lauffenburger DA, Zhang S, Kamm RD (2002) Left-Handed Helical Ribbon Intermediates in the Self-Assembly of a β-Sheet Peptide. Nano Letters 2:295–299.

Nagai A, Nagai Y, Qu H, Zhang S (2007) Self-assembling behaviors of lipid-like peptides $A^6D$ and $A^6K$. J. Nanoscience & Nanotechnology 7:2246–2252.

Narmoneva DA, Oni O, Sieminski AL, Zhang S, Gertler JP, Kamm RD, Lee RT (2005) Self-assembling short oligopeptides and the promotion of angiogenesis. Biomaterials 26:4837–4846.

Oliver C, Waters JF, Tolbert CL, Kleinman HK (1987) Culture of parotid acinar cells on a reconstituted basement membrane substratum. J Dent Res 66:594–595.

Palsson B, Hubell J, Plonsey R, Bronzino JD (2003) Tissue engineering: Principles and applications in engineering. CRC Press: Boca Raton, FL

Pauling L (1960) The Nature of the Chemical Bond Third Edition. Cornell University Press: Ithaca, NY

Ratner B, Hoffman A, Schoen F, Lemons J (1996) Biomaterials Science. Academic Press: New York

Rich A, Davies D (1956) A new two-stranded helical structure: polyadenylic acid and polyuridylic acid. J Amer Chem Soc 78:3548.
Santoso S, Hwang W, Hartman H, Zhang S (2002) Self-assembly of surfactant-like peptides with variable glycine tails to form nanotubes and nanovesicles. Nano Letters 2:687–691.
Schmeichel KL, Bissell MJ (2003) Modeling tissue-specific signaling and organ function in three dimensions. J Cell Sci 116:2377–2388.
Timpl R, Rohde H, Robey PG, Rennard SI, Foidart JM, Martin GR (1979) Laminin--a glycoprotein from basement membranes. J Biol Chem 254:9933–9937.
Vauthey S, Santoso S, Gong H, Watson N, Zhang S (2002) Molecular self-assembly of surfactant-like peptides to form nanotubes and nanovesicles. Proc Natl Acad Sci USA 99:5355–5360.
von Maltzahn G, Vauthey S, Santoso S, Zhang S (2003) Positively charged surfactant-like peptides self-assemble into nanostructures. Langmuir 19:4332–4337.
Weaver VM, Howlett AR, Langton-Webster B, Petersen OW, Bissell MJ (1995) The development of a functionally relevant cell culture model of progressive human breast cancer. Semin Cancer Biol 6:175–184.
Yannas IV (2001) Tissue and organ regeneration in adults.
Yang S, Zhang S (2006) Self-assembling behavior of designer lipid-like peptides. Supramolecular Chemistry 18:389–396.
Yaghmur A, Laggner P, Zhang S, Rappolt M (2007) Tuning curvature and stability of monoolein bilayers by designer lipid-like peptide surfactants. PLoS ONE 2:e479.
Yeh JI, Du S, Tordajada A, Paulo J, Zhang S (2005) Peptergent: peptide detergents that improve stability and functionality of a membrane protein glycerol-3-phosphate dehydrogenase. Biochemistry 44:16912–16919.
Yokoi H, Kinoshita T, Zhang S (2005) Dynamic reassembly of peptide RADA16 nanofiber scaffold. Proc Natl Acad Sci U S A 102:8414–8419.
Zhang S (2002) Emerging biological materials through molecular self-assembly Biotechnology Advances 20:321–339.
Zhang S, Altman M (1999) Peptide self-assembly in functional polymer science and engineering. Reactive & Functional Polymers 41:91–102.
Zhang S, Lockshin C, Herbert A, Winter E, Rich A (1992) Zuotin, a putative Z-DNA binding protein in Saccharomyces cerevisiae. EMBO J 11:3787–3796.
Zhang S, Holmes T, Lockshin C, Rich A (1993) Spontaneous assembly of a self-complementary oligopeptide to form a stable macroscopic membrane. Proc Natl Acad Sci U S A 90:3334–3338.
Zhang S, Lockshin C, Cook R, Rich A (1994) Unusually stable beta-sheet formation in an ionic self-complementary oligopeptide. Biopolymers 34:663–672.
Zhang S, Holmes TC, DiPersio CM, Hynes RO, Su X, Rich A (1995) Self-complementary oligopeptide matrices support mammalian cell attachment. Biomaterials 16:1385–1393.
Zhang S, Zhao X, Spirio L (2005) PuraMatrix: Self-assembling peptide nanofiber scaffolds. In Scaffolding in Tissue Engineering. (Ed. Ma & Elisseeff) CRC Press, Boca Raton, FL pp.217–238.
Zhao X, Zhang S (2006) Self-assembling nanopeptides become a new type of biomaterial. Adv Polym Sci 203:145–170.
Zhao X, Nagai Y, Revees P, Kiley P, Khorana HG, Zhang S (2006) Designer lipid-like peptides significantly stabilize G-protein coupled receptor bovine rhodopsin. Proc Natl Acad Sci USA 103:17707–17712.
Zhau HE, Goodwin TJ, Chang SM, Baker TL, Chung LW (1997) Establishment of a three-dimensional human prostate organoid coculture under microgravity-simulated conditions: evaluation of androgen-induced growth and PSA expression. In Vitro Cell Dev Biol Anim 33:375–380.

# Part IV
# Nanomedicine: Nanotechnology for Diagnosis and Treatment

# Quantum Dot Imaging of Neural Cells and Tissue

**Tania Q. Vu and Sujata Sundara Rajan**

## 1 Introduction

Nanomedicine is an emerging field of science and technology that integrates biological, chemical, and engineering sciences and holds promise for the creation of nanometer-sized scale approaches for studying biological systems in health and disease. While biologists, biochemists, and biophysicists have long been interested in biomolecular phenomena and cellular ultrastructure at the nanometer scale, advances in atomic scale visualization technology in the past few decades have motivated focused interest in understanding and manipulating synthetic and biomaterials at the molecular level. Nanomedicine, the application of nanotechnology to solving health problems, seeks to establish tools and techniques that provide easy accessibility to the interior of cells, facilitates molecular scale interaction with cellular signaling cascades, and increases detection sensitivity and targeting at the single-molecule level. Through the development of materials that exhibit novel optical, chemical, and electrical properties at the nanometer-sized scale, it is hoped that it will be possible to probe, study, and transform molecular cell processes to further understand fundamental biological processes and to improve therapies for disease.

The field of nanoparticle synthesis, assembly, and application to biology is a fast growing area of nanotechnology and nanomedicine (Niemeyer 2001; Penn et al. 2003; Salata 2004). Nanoparticles are one of the earliest and most prevalent areas of nanomedicine research and work in this area is entering a period of rapid investigation along with significant commercial interest for development of therapeutics (West and Halas 2003; Salata 2004). Nanoparticle spheres, rods, and cylinders (1–100 nm) measure on the same size scale as proteins and other biomolecules and possess high surface-to-volume ratios. These two traits imply that nanoparticles

T.Q. Vu (✉) • S. Sundara Rajan
Department of Biomedical Engineering, Oregon Health & Sciences University, Portland, OR, USA
e-mail: vuta@ohsu.edu

G.A. Silva and V. Parpura (eds.), *Nanotechnology for Biology and Medicine: At the Building Block Level*, Fundamental Biomedical Technologies,
DOI 10.1007/978-0-387-31296-5_7, 

could serve as useful nanoscale platforms to target, probes, and deliver chemicals to specific subcellular structures.

A unique class of nanoparticles is quantum dots (QDs). QDs are fluorescent spherical nanoparticles that have extremely well-controlled size dimensions and exhibit novel photophysical properties, including extended photostability and multicolor excitation. The unique physical properties of QDs have only begun to be understood and their application to biology is very recent. However, even at these early stages, it is clear that QDs offer capabilities that will make significant future contributions to biology. In the past few years, there has been rapid progress along the lines of QD synthesis and surface chemical developments. As the properties of QDs and their interaction in physiological systems become more clearly understood, QD-based biological applications are poised for further development of more sophisticated and specialized biological and disease-oriented problems.

The nervous system is a complex organ that presents unique challenges for study at the systems, tissue, and cellular level. Knowledge about the fundamental central and peripheral neural processes underlying neuronal signaling, development, learning, and sensing have been enabled by crucial technological advances in physics, chemistry, and engineering. We believe that QD nanoparticles can further enable significant technological advances in neuroscience. In this chapter we review the history and development of QDs and their initial application in biology, specifically concentrating on current QD-neuroscience applications. We also discuss specific areas of neuroscience that would benefit from development of future QD applications. QDs are largely used in biological studies as substitutes for fluorescent tags, but they are yet to be fully developed to address a variety of specific and significant questions in neuroscience basic research and biomedical therapies. The successful results of QD-based work in a broad range of cell systems and applications indicate that the convergence of development of QDs for neuroscience will prove to be a rewarding area of future technology advancement.

## 2 Unique Challenges Presented by Neuroscience

The problem of trying to understand the brain is an immense and demanding challenge. The brain is the most complex tissue in the body, with a greater diversity of cell types than all other organs and tissues combined. Furthermore, 100 billion of neurons each make synaptic connections to one another, processing and propagating dynamic information that serves as the underlying mechanism of behavior, cognition, neural development, learning, and sensory and autonomic processing. The nanometer size and complexity of neuroanatomical structures, small quantities of neurochemicals, rapid time course of exchange, and dynamic flexibility of this process present significant challenges for understanding fundamental neural function. Neurotransmission is mediated by the release of neurotransmitter ligands from as little as a one vesicle, a spherical structure 30–50 nm in diameter, into a synaptic cleft measuring 30 nm in width (Rieke and Schwartz 1996; Zhai and Bellen 2004;

Heidelberger et al. 2005) at rapid timescales (<1 ms) (Burgoyne and Barclay 2002; Heidelberger et al. 2005). The interplay between ligand binding to receptor channels is a dynamic process, with rapid insertion and clearance of receptors from subcellular synaptic domains occurring on millisecond time scales (Triller and Choquet 2003; Thomas et al. 2005). Ligand–receptor binding events produce downstream modification of postsynaptic subcellular constituents in highly confined calcium ion microdomains, enclosed nanovesicles, and along cytoskeletal transport elements. Moreover, in neural tissue, the complex connectivity of the different types of neurons makes it difficult to identify and distinguish specific synapses for study. In vivo imaging and drug targeting of brain structures present an additional challenge of bypassing the blood–brain barrier (BBB). Thus, highly sensitive, rapid, cell-accessible and cell-specific means are needed to detect and target chemical events occurring in nanometer-sized subcellular structures.

## 3 Historical Development of Quantum Dots

Nanocrystals, nanoparticles with crystalline cores, can be synthesized with precise shape and sizes (1–10 nm) using wet-chemical methods. As a general rule, the optical, electrical, and magnetic properties of these nanocrystals are largely determined by their shape and size and can thus be tailored by their dimensions (Vanmaekelbergh and Liljeroth 2005). Presently, technological interest in nanoparticles stems from the powerful implication that novel materials with distinct physical properties can be created by controlling the size and shape of nanoparticles during synthesis. Thus, nanocrystals can serve as versatile building blocks for designing nanoscale devices with desired physical properties.

Colloidal nanocrystals can be synthesized from noble metals, transition metals, and many semiconductor compounds (Murray et al. 2000). QDs are single spherical nanocrystals (1–100 nm) that are made from semiconductor materials. Semiconductor QD research was originally motivated by understanding electronic structure on a molecular scale and the use of these materials to develop new opto-electronic solid state devices. Recently, it has been realized that the size and unique optical properties of QDs also have useful implications in biology and medicine. In 1998, initial studies using QDs for biological labeling were reported (Bruchez et al. 1998; Chan and Nie 1998) and since then have generated an exponential growth in the development of QD-based biomedical applications (Pinaud et al. 2006).

## 4 Physical Properties of Quantum Dots

The unique size and semiconductor composition of QDs confer upon them unusual photophysical properties. The size of a QD is comparable to that of the bulk Bohr radius (~56 Ä for CdSe). This results in confined movement of free charge carriers,

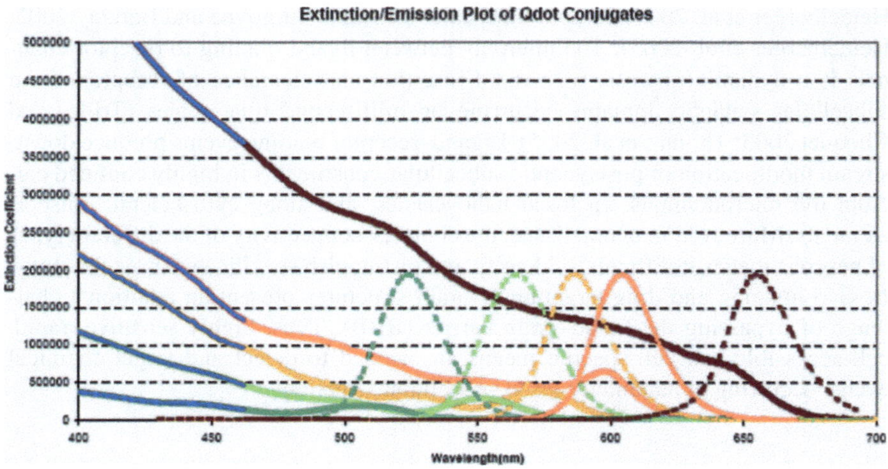


**Fig. 1** Characteristic absorbance and emission spectra for a series of five different quantum dot sizes. Absorbance spectra are characterized by broad band excitation that most efficiently excites at lower wavelengths (denoted by the *blue* portion of absorbance curves). Emission spectra are characterized by symmetric, narrow Gaussian-shaped curves, which shift to higher wavelengths with increasing quantum dot size. *Source*: Quantum Dot Corp, http://www.qdot.corp

electrons, and holes, and QDs, unlike their bulk semiconductor materials, possess quantized energy levels (Murray et al. 2000; Parak et al. 2003; Michalet et al. 2005). These energy levels increase with decreasing QD size and QDs and by varying the size and chemical core composition such that QDs' luminescence spans across the visible spectrum (Nirmal and Brus 1999). Another consequence of the QD size being smaller than the Bohr radii for the material is that the semiconductor material now has a greater probability of absorbing light at higher energies (shorter wavelength) and emitting light at high energies (shorter wavelength). Figure 1 shows a typical QD absorption and emission spectrum. Absorption of a photon with energy above the semiconductor band gap energy, the energy difference between the top of the valence band and the bottom of the conduction band, results in the creation of an electron–hole pair that has a broadband absorption spectrum with increased probability at shorter wavelengths (Michalet et al. 2005). As the QD size decreases, electron–hole pairs experience stronger confinement and produce shorter wavelength (higher energies) when photoexcited. The radiative recombination of electron–hole pairs produces an emission of a photon that possesses a narrow and symmetric energy band. Thus excitation spectra track that of absorbance spectra and will emit with the same wavelength spectrum independent of the excitation wavelength (Bruchez et al. 1998). Taken together, semiconductor QDs possess a unique absorbance and emission spectrum as a physical consequence of their nanometer size. Currently, QDs are produced from a variety of materials such as: cadmium sulfide, cadmium selenide, cadmium telluride (periodic groups II–VI), or gallium arsenide and indium phosphide (III–V), or lead sulfide, lead selenide, silicon, and germanium

(IV–VI) (Alivisatos et al. 2005; Vanmaekelbergh and Liljeroth 2005). Initially, QDs made from CdSe are of particular interest because when excited optically, by varying the QD size and chemical composition, the wavelength of the fluorescence light emitted can be tuned to a broad emission spectrum that includes visible light (400–2,000 nm) (Murray et al. 2000; Fu et al. 2005). CdSe QDs were produced in 1993 using high-temperature organometallic chemical techniques (Goldstein et al. 1992; Peng and Peng 2001). This allowed for monodispersed QDs with narrow size variations (SD < 5%) (Murray et al. 2000). Further development in depositing a surface capping layer of ZnS or CdS increased quantum yields from ~10% to ~40–50% (Chan et al. 2002).

## 5 Quantum Dot Optical Properties Expand Bioimaging Capabilities

Optical detection is perhaps the most widely used modality for detecting and imaging biological systems. Over the past decade, rapidly increasing number of fluorescent dyes and proteins have been developed, resulting in improved ability to image subcellular processes with higher resolution and enabled major advances in neuroscience (Miyawaki et al. 2003). Fluorescent QDs exhibit unique photophysical properties that are dramatically different from organic dyes. These properties have been demonstrated in an increasing number of diverse cellular systems and preparations, thus expanding the biologist's toolbox of capabilities that have not been possible with existing techniques.

QD fluorescence is exceedingly bright and photostable compared to conventional organic dyes. Brightness can be approximated by the product of quantum yield and extinction coefficient at the excitation wavelength. Although QDs possess quantum yields that are comparable to that of organic dyes, their molar extinction coefficients ($10^5$–$10^6$ $M^{-1} cm^{-1}$) far exceed that of organic dyes by ×10–100 (http://www.qdots.com/live/render/content.asp?id=84; Bruchez et al. 1998; Chan et al. 2002). QD brightness has been calculated to be about ~20 times brighter than that of single organic dye molecules such as rhodamine (Chan et al. 2002). Due to their inorganic composition, QD fluorescence is also exceptionally photostable compared to existing organic fluorophores (Alivisatos et al. 2005). For example, Jaiswal et al. showed that under continuous 50-mW illumination QDs did not photobleach even after 14 h. In contrast, most organic fluorophores are subject to fast photobleaching, completely losing their ability to emit in less than 20 min (Jaiswal and Simon 2004). In in vitro and in vivo studies, QDs have proven to not undergo any significant bleaching and have enabled monitoring of molecules in live cells for hours, days, weeks, and months (Jaiswal and Simon 2004). Such unprecedented QD brightness and photostability means that cellular structures can be detected at high sensitivity over extended periods of time without significant reduction in intensity. Indeed, it has been reported that a very small number of individual QDs are needed to immunolabel subcellular receptors and may in fact approach that of one QD per target molecule.

This increased sensitivity, combined with unprecedented photostability, has allowed for single proteins to be dynamically tracked in live cells studies (Dahan et al. 2003; Lidke et al. 2004).

QDs possess a broad excitation that favors excitation at low (blue) visible wavelengths. Unlike conventional organic dyes, the QD emission spectrum is symmetrical and quite narrow (30–40 nm) and can be precisely tuned by varying QD size (Bruchez et al. 1998; Michalet et al. 2001). Furthermore, the Stoke's shift, the separation between excitation and emission wavelengths, of QDs is large compared to that of organic dyes and further enhances the ability for multicolor imaging. Altogether, this allows QDs to serve as fluorescent tags for a wide spectrum of colors, which can be excited with a single wavelength or excitation source and can be viewed simultaneously. Indeed, QD barcode technology has demonstrated that by multiplexing 5–6 QD colors at six intensity levels in polymer beads, it would be possible to encode among thousands of multiple emissions colors (Han et al. 2001). While any one of the above QD photophysical properties would alone be a marked improvement over conventional organic dyes, taken together they represent a dramatic leap forward in fluorescent imaging technology.

There are some new photophysical properties of QDs that are not well understood and have not been well examined in cellular applications. For example, QDs undergo fluorescence intermittency, or blinking, caused by surface defects in the nanocrystal, which act as traps for electron–hole pairs and prevent their recombination (Nirmal and Brus 1999; Yao et al. 2005). Blinking "on" durations vary inversely with excitation intensity while the "off" state is excitation independent and is dependent on the QD coating and is present for QDs immobilized onto surfaces as well as in aqueous physiological environments (Nirmal and Brus 1999; Yao et al. 2005). Remarkably, the blinking lasts on a several second time scale (0.1 s to ~10 min) (Yao et al. 2005). On a practical basis, blinking is a convenient signature of single QDs. Since aggregates of QDs will cancel out this behavior through statistical coincidence, the presence of intermittent fluorescence in QD samples can serve as a good indicator of single molecule labeling. However, because QD blinking can vary in rate and is dependent on surrounding environment, if not well controlled, can prove to be a problematic artifact of QD fluorescence and interfere with applications such as estimating dynamic single-molecular tracking at rapid timescales as well as fluorescence resonant energy transfer studies. In addition, blinking is correlated with spectral jumping, or change in emission peak position (Pinaud et al. 2006). Currently, reduction in blinking is being addressed by changing the composition of the core and thickening of the outer shell, which it is believed will eliminate blinking (Michalet et al. 2005). The parameters that influence blinking in physiological environments are just only beginning to be investigated. In the future, if blinking can be well-controlled, this property may actually prove useful to sense localized environmental conditions.

Another less investigated photophysical property of QDs is the influence of local physiological environments on QD fluorescence. QD fluorescence is highly dependent on environmental conditions. Spin-coated QDs exhibit changes in fluorescence fluctuations when exposed to dry versus humid nitrogen environments (Michalet et al. 2001) and QD fluorescence intensity increases upon photoexcitation, possibly caused by local environmental interactions with the QD surface. Moreover, Silver and Ou have looked at QD fluorescence of endocytosed QDs and have observed that intracellular QDs exhibited "photoactivation" or "photobrightening" after ~1 min of photoactivation (Silver and Ou 2005). QDs that were barely detectable in lightly labeled intracellular structures became markedly brighter and were observed more strongly in cells in PBS than in ethanol. While environment-dependent variations could prove problematic for QD fluorescence quantification and tagging, further studies to carefully examine the effect of pH, lysosomal degradation of specific QD surface molecules, and other local environmental intracellular factors could provide new opportunities for development of environmentally sensitive high resolution cellular probes.

## 6 Quantum Dot Surface Chemistry for Cellular Interaction

Surface chemistry that allows tethering of bioactive molecules to the QD surface is a crucial factor in customizing QDs for specific biological interactions in cells. QDs that have hydrophobic surfaces are not directly soluble in aqueous solution and may aggregate in the presence of physiologically relevant electrolytic concentrations. A challenge for adapting QDs for biological applications has been to find suitable surface coatings that maintain reproducible photophysical properties, stability in aqueous solution, and chemical conjugation/manipulation in physiological conditions. An additional critical factor to consider when designing QD surface coatings is to maintain the total end diameter so that increase in QD size does not negatively impact the mobility and accessibility of the resulting QD-conjugated biomolecules (Jaiswal and Simon 2004). Surprisingly rapid progress has been made in finding suitable coatings for functionalizing QD surfaces to meet these above criteria in the past few years. Initial QD coatings included mercaptoacetic acid, dihydrolipoic acid, or modified amphiphilic polymers such as polyacrylic acid. These approaches offer the advantage of serving as very general and can be adopted for other nanomaterials with similar hydrophobic surfactants on their surface (Chan and Nie 1998; Goldman et al. 2002; Jaiswal et al. 2003; Wu et al. 2003; Pinaud et al. 2006). Another general approach has been to use exchange chemistries such as di-thiol ligands with additional coatings of engineered proteins and peptides to cross-link the ligands and thus provide a more stable coating (Pinaud et al. 2004, 2006).

Once QDs have hydrophilic coatings, strategies must be devised to bind ligands and other biomolecules of interest. To date, biomolecules have been linked to water-dispersible QDs using covalent attachment via –COOH, –SH, or $NH_2$ groups, electrostatic attraction (Goldman et al. 2002), and biomolecular protein linking systems (Fig. 2). The widely used avidin–biotin system has also been used to link immunoglobulin G molecules to QD surfaces (Wu et al. 2003). Biomolecules such as antibodies, biotin, oligonucleotides, peptides, and proteins have been bound to QD surfaces using these linking systems. In further developments, Howarth et al. have used the *Escherichia coli* enzyme biotin ligase-acceptor peptide system to label

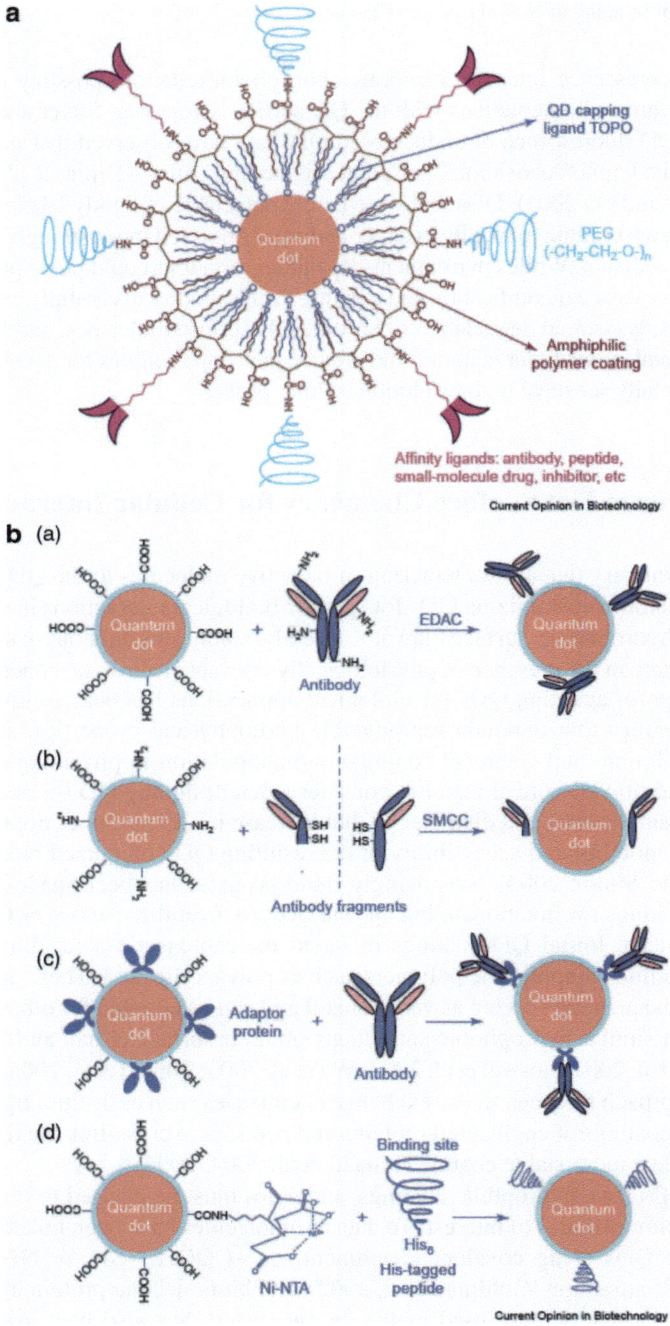


**Fig. 2** (**a**) Schematic of a multifunctional quantum dot probe showing the capping ligand TOPO and an outer co-polymer layer, which could be coupled to affinity ligands or PEG. (**b**) Conjugation of QDs to biomolecules (a) through covalent cross-linking using EDAC (b) binding anti-body fragments via reduced sulfhydryl-amine coupling with SMCC (c) by means of an adaptor protein (d) conjugation of histidine-tagged peptides to Ni-NTA modified QDs. *Source*: Gao et al. (2005)

glutamate-activated α-amino-3-hydroxy-5-methyl-isoxazole propionate (AMPA) channels/receptors with good specificity (Howarth et al. 2005). Pinaud et al. have designed a synthetic cysteine-rich phytochelatin-related peptide and used it to coat QDs (Pinaud et al. 2004). QDs coated with peptides containing glutamic acid residues showed good solubillization, stability and unaltered photophysical properties. The authors also showed successful labeling of cells expressing avidin-CD14 chimeric proteins with biotinylated peptide coated QDs. It is to be expected that future surface modifications using alternative organic QD coatings such as these peptides will improve the size and stability of QDs for biological applications.

## 7 Quantum Dot Toxicity

Due to their heavy metal cadmium and selenium composition, there have been concerns related to the potential toxicity of QDs. Currently, most reports have not found evidence for QD cytoxicity as assessed by change in cell morphology in in vitro cell cultures (Jaiswal et al. 2003), and by injecting them into embryos and other animals models for up to 4 months (Dubertret et al. 2002; Alivisatos et al. 2005). The release of $Cd^{2+}$, enhanced by UV surface oxidation of QDs, has been reported to harm cultured cells, but this effect has been shown to be significantly repressed by the formation of a ZnS shell around the QD CdSe core (Derfus et al. 2004b). Other sources of cytotoxicity could originate not from release of $Cd^{2+}$ but rather from the biomolecular QD surface coatings or from potential aggregation that surface coating may produce (Hoshino et al. 2004; Kirchner et al. 2005). It has also been proposed that the size of QDs, once they get endocytosed, could interfere with normal cell function, and produce toxic effects though this has not yet been thoroughly investigated (Shiohara et al. 2004). The potential harmful effects of QDs will depend on parameters that differ from study to study, such as specific chemical composition of the QD core and shell (ZnS shell, polyethylene glycol coat), as well as environmental conditions, such as illumination, time scale of exposure, and quantities exposed. More studies that use more stringent toxicological criteria will be needed to examine factors that may bring about QD longer periods of time.

## 8 Current Applications of Quantum Dots to Neuroscience

### *8.1 Quantum Dots for Immunolabeling Neural Receptors*

Receptor composition, distribution, and dynamics play a central role in initiating various signaling pathways in the nervous system mediated by any single type of particular ligand, or chemical group such as a neurotransmitter, antigen, hormone or synthetic drug analog (Gur and Yarden 2004). The composition of receptors on any neuron varies in population and density as a function of space and time. Commonly used techniques such as radiolabeling or immunohistochemical techniques suffer from the drawbacks of limited spatial-temporal resolution and lack of real time monitoring. Fluorescence imaging techniques currently offer the most information about receptor dynamics; however, organic dyes provide only momentary visualization of receptor distribution and movement due to rapid photobleaching effects ($t < 10$ s). Moreover, photostable fluorescent probes such as green fluorescent protein (GFP) are poor fluorophores for single molecule imaging as it is difficult to determine if the fluorescent signal is coming from the cell surface, the cell interior or a combination of both these sites (Niemeyer 2001; Vu et al. 2005).

Early applications of QDs to study receptor dynamics suggest that they will be a significant improvement over current fluorescent tags. Dahan et al. used anti-glycine antibody-conjugated QDs and confocal microscopy to track individual glycine receptors in cultured spinal cord neurons (Dahan et al. 2003). The small size of the antibody-conjugated QDs enabled access to the receptors in the synaptic cleft and provided a significant improvement over latex beads (200–500 nm in diameter) and colloidal gold (40 nm in diameter) previously used to study receptor dynamics. The relatively extreme brightness and photostability of the QD fluorescence enabled the use of these probes for tracking receptors in live cells over an extended time course of 20 min. In addition, the blinking of these dots confirmed the detection of individual receptors and this property of QDs was used to follow receptor movement using time-lapse imaging. The fluorescence tracking study helped distinguish glycine receptors into three membrane sub-domains based on their distinct diffusion properties.

Receptor mapping has also been demonstrated by Howarth et al. (Howarth et al. 2005). A special feature of their study is that they observed glycine-evoked changes in QD-tagged AMPA receptor distribution in hippocampal neurons transfected with GluR1 or GluR2 subunits. GluR1 receptor sub-units were barely detectable before the addition of glycine, but a significant population was observed after application of the glycine pulse. On the other hand, the GluR2 receptors showed little change in distribution in response to glycine. Interestingly, they observed that the size of QDs could limit their accessibility to a portion of GluR2 subunits. QD-tagged receptors were detected co-localized with a chimeric post-synaptic density protein PSD95-yellow fluorescent protein (YFP). However, in experiments where the AMPA receptor subunits were labeled with Alexa Fluor™ dyes instead of QDs, greater co-localization with PSD95-YFP was found, than in the case of QD labeling.

In the future, new experiments using QD-tagged receptors will continue to contribute to new knowledge concerning receptor spatial-temporal dynamics. In addition, receptor mapping using multiple QD colors will be useful for understanding the interactions of multiple types of receptors and other synaptic proteins in live cell preparations.

## 8.2 *Quantum Dots as Ligand–Receptor Probes*

Presynaptic neurons release ligands that act on the receptors of postsynaptic neurons at close proximity as well as over longer distances. Where do these ligands go, which subtype of receptors do they bind to, and how effective are they in binding to postsynaptic receptors? Traditional radiolabelling assays provide good sensitivity and quantitative information on ligand–receptor binding; however radioactive probes lack spatial and temporal information. Fluorescent or nanoparticle tagged ligands would provide spatial information; however, at present the availability of such probes is limited. The variety of QD bioconjugation methods offer good potential for designing neuroactive ligand-conjugated QD probes to study receptor-mediated activity and track drug binding and uptake in pharmaceutical and therapeutic questions.

Initial studies using ligand-conjugated probes have shown that these probes bind to their specific receptors without exhibiting significant steric hindrance and provide high spatial resolution. Lidke et al. conjugated QDs to epidermal growth factor (EGF) and monitored its binding to erbB1 receptors in CHO cells (Lidke et al. 2004). Appending these receptors with enhanced GFP (eGFP) enabled simultaneous tracking of the ligand–receptor complex and its endocytosis into the cell. Dynamic cellular processes, such as membrane ruffling, were observed prior to the uptake of EGF-QD/erbB1-eGFP complexes and retrograde transport of these receptors was observed in the filopodia of A431 cells.

The binding of nerve growth factor (NGF) to QDs to activate TrkA receptors in PC-12 cells was studied by Vu et al. (2005). Short-term exposure to beta-NGF conjugated QDs, revealed specific QD binding to TrkA receptors, followed by internalization of the ligand–receptor complexes. Long-term studies (3–5 days) demonstrated that the NGF-QDs after binding to TrkA receptors, remained bioactive and were capable of mimicking NGF activity as exhibited by extensive neurite sprouting.

Rosenthal et al. used QDs to gain more insight into transporter localization and regulation (Rosenthal et al. 2002; Tomlinson et al. 2005). They showed that serotonin-conjugated QDs were not effective in inducing a current from the serotonin-3 receptors but were able to block the serotonin-transporter by binding to a site which either overlaps with an antagonist site or is the serotonin-binding site itself. The serotonin-QDs enabled direct visualization of the serotonin transporter protein on the surface of transfected HEK cells and staining was completely eliminated on preincubation with an antagonist.

These early studies demonstrate that QDs can be used as sensitive probes to visualize ligand/drug trafficking and study the signaling pathways of ligands of interest. These results are encouraging as they show that QDs can be developed as new tools to track efficacy of ligand–receptor binding and the fate of ligand–receptor complexes.

## 8.3 *Concurrent Quantum Dots Fluorescence and Electron Microscopy Imaging of Neural Ultrastructures*

An advantage of QD probes is that they can be used as concurrent fluorescent and electron dense electron microscopy (EM) probes, thus giving access to additional information in the same biological preparation. Dahan et al. performed single-receptor

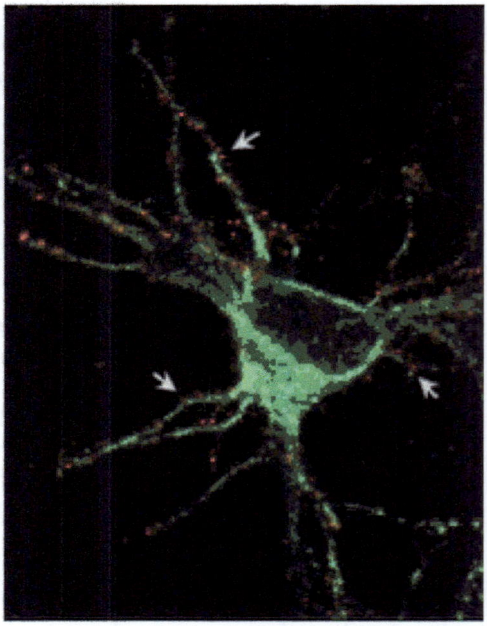

**Fig. 3** QDs (*red*) used to label Glycine receptors on cultured spinal cord neurons (*arrows*). The somatodendritic compartment is identified by microtubule associated protein-2 (*green*). *Source*: Dahan et al. (2003)

tracking of glycine receptors using confocal fluorescence imaging, then used silver intensified QDs examined with EM to confirm the location of the antibody-QD tagged glycine receptors in the synaptic cleft (Dahan et al. 2003) (Fig. 3). Thus the same QDs probes that were used to track these receptors in the neuronal membrane using fluorescence were also imaged with electron microscopy, providing high resolution information. In a careful study using correlated light microscope and electron microscope labeling, Giepmans et al. report successful labeling of three proteins at both the light and EM level within the same tissue sample (Giepmans et al. 2005). Although postfixation of the sample with glutaraldehyde and osmium was eliminated in order to prevent loss in fluorescence, the ultrastructure of the cells was sufficiently intact for identifying detail.

# 9 Future Application of Quantum Dots to Neuroscience

## *9.1 Quantum Dots Imaging for Tracing Single Neurons and Subcellular Organelles*

While immunolabeling of neuronal structures has been immensely useful in elucidating neural function, antibody assays do not allow for distinction of individual neuronal structures such as axons and dendrites from that of neighboring cells in neural tissue. In addition, immunolabeling techniques are often performed in fixed tissue, while the ability to labeling and monitor neuronal structures in live tissue would yield more information in dynamic situations. Intracellular dyes, such as Lucifer Yellow, carbocyanine dyes such as DiI, DiO, and DiD, or neurobiotin, have been delivered by microinjection as well as ballistic labeling and have successfully been used to fill target neurons. Although this allows for selective labeling in live neural in vitro preparations; toxicity and photobleaching continue to limit the usefulness of these dyes (Mobbs et al. 1994; Morgan et al. 2005). Although QDs have not yet been used as neuronal tracers, this area of research may yield fruitful results. The potential to modify size, charge, and other QD surface properties could yield a new class of neural tracers that are brighter, more mobile, and can interact with specific intracellular constituents to track intracellular neural events.

A current challenge in using QDs for visualizing and monitoring subcellular structures is delivery of QDs to the cellular organelles of interest. The size of these QDs precludes diffusion through the plasma membrane, unlike organic dyes, which are used to label intracellular organelles. Derfus et al. showed that intracellular delivery of QDs to cells by incubation of QDs conjugated to PEG and peptide transfection agents as well as using electroporation have successfully resulted in QD delivery to the cellular cytosol without being captured in endosomes (Derfus et al. 2004a). A drawback to both these methods, however, is that they produce QD aggregates of up to 500 nm in diameter. Alternatively, microinjected delivery of QDs complexed with PEG and a nuclear or mitochondrial localization sequence produced specific labeling and the absence of QD aggregates (Derfus et al. 2004a).These results indicate that it is possible to deliver single QDs to the cellular cytosol.

In the future, by successful pairing of a chosen QD delivery technique with specific surface-conjugated biomolecules, it may be possible to not only prevent aggregation of QDs but also to control interaction of QDs with specific intracellular neural constituents. By conjugating QD surfaces with suitable proteins or peptides, it is conceivable that QDs can be developed as quantitative neuroactive probes. For example, by designing QDs that could bind to specific synaptic vesicle proteins, it may be possible to track neural activity with lower toxicity or high background fluorescence associated with existing styryl dyes such as FM-143 (Brumback et al. 2004).

## 9.2 *Quantum Dots for Bioimaging Brain and Retinal Function*

Potential QD imaging techniques may yield exciting future research directions for studying brain and retinal function in intact, live tissue. QDs have efficient multiphoton excitation cross sections and can emit infrared or near-infrared light and thus they are compatible with multiphoton imaging techniques used for in vivo imaging of cells in thick tissue sections (Larson et al. 2003; Kim et al. 2004; Ballou et al. 2005). In vivo imaging studies in a variety of biological systems have shown a number of promising results. Larson et al. intravenously injected water-soluble QDs in the tail vein of mice and showed that these probes can be used to successfully image skin and adipose tissue (Larson et al. 2003). Comparison of QDs to fluorescein isothiocyanate-conjugated dextran imaged at the same depth showed that QD probes were brighter and hence revealed greater detail. Near-infrared QDs have been injected intradermally in a pig and imaged real-time with virtually no autofluorescence of the tissue (Kim et al. 2004). This technique has allowed surgeons to follow the migration of the QDs towards the sentinel lymph node (SLN) and identify the position of the SLN quickly, providing visualization and mapping during surgical procedures, thus eliminating the need for both the radioactive tracers (Kim et al. 2004).

The BBB tightly regulates the passage of selected molecules into the brain parenchyma and thus limits the accessibility of biomolecules to the brain presenting a challenge for the in vivo delivery of imaging agents and compounds to the brain. Accumulating evidence in the past 2 decades has shown, however, that proteins and peptides are capable of crossing the BBB via specific transport systems such as receptor and carrier-mediated endothelial cell transport systems (Pan and Kastin 2004; Costantino et al. 2005; Olivier 2005). Recent work using polymeric nanoparticles (10–100 nm) with surface-conjugated antibodies and other peptides has shown success in crossing the BBB (Roney et al. 2005). Although QD delivery to the brain has yet to be investigated, the small size of QDs as well as their ease of conjugation to a variety of compounds makes them promising candidates for similar passage across the BBB. Their property of fluorescence eliminates the need for conjugation with a dye molecule as in the case of polymeric nanoparticles and thus QDs can serve as probes for labeling of neural targets in live brain tissue as well as multi-effector platforms for drug transport across the BBB.

Nanoparticles have been delivered into the eye with greater ease than the brain due to better accessibility and may prove to be a feasible first step to imaging intact neural tissue as well as for further development of nanoparticle therapies. In their study, de Kozak et al. injected fluorescent Poly (methoxypolyethyleneglycol cyanoacrylate-co-hexadecyl cyanoacrylate) (PEG-PHDCA) nanoparticles into the vitreous of rats with experimental autoimmune uveoretinitis and found that the nanoparticles dispersed freely in the anterior and posterior sections of the eye and infiltrated the retina through the end feet of glial cells (de Kozak et al. 2004). When the nanoparticles were loaded with tamoxifen, the disease severity was inhibited indicating the effectiveness of the intraocular treatment. Bourges et al. showed that polymer nanoparticles injected into the vitreous were mobile, entering the retina and retinal epithelium, while maintaining anatomical integrity of the tissue (Bourges et al. 2003).

## 10 Conclusions and Future Directions

QD-based biological applications have emerged only in the past few years; however, the rapid success of QDs for imaging cells and tissue in a variety of cellular systems and imaging modalities indicates the versatility and potential of this nanotechnology to contribute to biomedicine. In the realm of neuroscience, the initial studies highlighted in this chapter demonstrate that the unique properties QDs are extremely well-suited for the highly sensitive, rapid, and nanometer scale resolution necessary for studying neural structures and subcellular receptor dynamics. These promising results have already catalyzed interest for further creative use of QDs tailored to address specific neuroscience questions. As understanding of QD properties and their interactions in physiological systems advances, QDs applications will move beyond substitutes for fluorescent tags, and will integrate other QD properties including high surface-to-volume ratio, size, and surface chemical interactions to tailored applications. Diagnostic and clinical applications using QDs in sensor devices and as drug delivery platforms is an underdeveloped area that will grow as QD-based biological applications and the commercial availability of QDs continue to expand.

Nanomedicine is a field in its infancy but is a rapidly growing field with demonstrable achievements. In just a number of years, QDs have transformed from a relatively obscure nanomaterial to one that is steadily achieving routinely use. The future challenge is to continue to understand the interactions of QDs in neural and other physiological systems and use this knowledge to engineer truly innovative tools to address challenging and unresolved biomedical problems.

**Acknowledgments** We thank Oregon ETIC funding and David Feigelson for his input in the initial stages of this manuscript.

## References

Alivisatos AP, Gu W, Larabell C (2005) Quantum dots as cellular probes. Annual Review of Biomedical Engineering 7:55–76, 53 plates.

Ballou B, Ernst LA, Waggoner AS (2005) Fluorescence imaging of tumors in vivo. Curr Med Chem 12:795–805.

Bourges JL, Gautier SE, Delie F, Bejjani RA, Jeanny JC, Gurny R, BenEzra D, Behar-Cohen FF (2003) Ocular drug delivery targeting the retina and retinal pigment epithelium using polylactide nanoparticles. Invest Ophthalmol Vis Sci 44:3562–3569.

Bruchez M, Jr., Moronne M, Gin P, Weiss S, Alivisatos AP (1998) Semiconductor nanocrystals as fluorescent biological labels. Science 281:2013–2016.

Brumback AC, Lieber JL, Angleson JK, Betz WJ (2004) Using FM1-43 to study neuropeptide granule dynamics and exocytosis. Methods 33:287–294.

Burgoyne RD, Barclay JW (2002) Splitting the quantum: regulation of quantal release during vesicle fusion. Trends Neurosci 25:176–178.

Chan WC, Nie S (1998) Quantum dot bioconjugates for ultrasensitive nonisotopic detection. Science 281:2016–2018.

Chan WC, Maxwell DJ, Gao X, Bailey RE, Han M, Nie S (2002) Luminescent quantum dots for multiplexed biological detection and imaging. Curr Opin Biotechnol 13:40–46.

Costantino L, Gandolfi F, Tosi G, Rivasi F, Vandelli MA, Forni F (2005) Peptide-derivatized biodegradable nanoparticles able to cross the blood-brain barrier. J Control Release 108:84–96. Epub 2005 Sep 2008.

Dahan M, Levi S, Luccardini C, Rostaing P, Riveau B, Triller A (2003) Diffusion dynamics of glycine receptors revealed by single-quantum dot tracking. Science 302:442–445.

de Kozak Y, Andrieux K, Villarroya H, Klein C, Thillaye-Goldenberg B, Naud MC, Garcia E, Couvreur P (2004) Intraocular injection of tamoxifen-loaded nanoparticles: a new treatment of experimental autoimmune uveoretinitis. Eur J Immunol 34:3702–3712.

Derfus AM, Chan WCW, Bhatia SN (2004a) Intracellular Delivery of Quantum Dots for Live Cell Labeling and Organelle Tracking. Advanced Materials 16:961–966.

Derfus AM, Chan WCW, Bhatia SN (2004b) Probing the Cytotoxicity of Semiconductor Quantum Dots. Nano Lett 4:11–18.

Dubertret B, Skourides P, Norris DJ, Noireaux V, Brivanlou AH, Libchaber A (2002) In vivo imaging of quantum dots encapsulated in phospholipid micelles. Science 298:1759–1762.

Fu A, Gu W, Larabell C, Alivisatos AP (2005) Semiconductor nanocrystals for biological imaging. Curr Opin Neurobiol 15:568–575.

Gao X, Yang L, Petros JA, Marshall FF, Simons JW, Nie S (2005) In vivo molecular and cellular imaging with Quantum Dots. Curr Opin Biotechnol 16:63–72.

Giepmans BN, Deerinck TJ, Smarr BL, Jones YZ, Ellisman MH (2005) Correlated light and electron microscopic imaging of multiple endogenous proteins using Quantum dots. Nat Methods 2:743–749.

Goldman ER, Balighian ED, Mattoussi H, Kuno MK, Mauro JM, Tran PT, Anderson GP (2002) Avidin: a natural bridge for quantum dot-antibody conjugates. J Am Chem Soc 124: 6378–6382.

Goldstein AN, Echer CM, Alivisatos AP (1992) Melting in Semiconductor Nanocrystals. Science 256:1425–1427.

Gur G, Yarden Y (2004) Enlightened receptor dynamics. Nat Biotechnol 22:169–170.

Han M, Gao X, Su JZ, Nie S (2001) Quantum-dot-tagged microbeads for multiplexed optical coding of biomolecules. Nat Biotechnol 19:631–635.

Heidelberger R, Thoreson WB, Witkovsky P (2005) Synaptic transmission at retinal ribbon synapses. Prog Retin Eye Res 24:682–720.

Hoshino A, Fujioka K, Oku T, Suga M, Sasaki YF, Ohta T, Yasuhara M, Suzuki K, K. Y (2004) Physiochemical properties and cellular toxicity of nanocrystal quantum dots depend on their surface modification. Nano Lett 4:2163–2169.

Howarth M, Takao K, Hayashi Y, Ting AY (2005) Targeting quantum dots to surface proteins in living cells with biotin ligase. Proc Natl Acad Sci USA 102:7583–7588.

Invitrogen Life Sciences, Qdot® Nanocrystal Technology Brochure: Quantum dot optical properties. In.

Jaiswal JK, Simon SM (2004) Potentials and pitfalls of fluorescent quantum dots for biological imaging. Trends Cell Biol 14:497–504.

Jaiswal JK, Mattoussi H, Mauro JM, Simon SM (2003) Long-term multiple color imaging of live cells using quantum dot bioconjugates. Nat Biotechnol 21:47–51. Epub 2002 Dec 2002.

Kim S, Lim YT, Soltesz EG, De Grand AM, Lee J, Nakayama A, Parker JA, Mihaljevic T, Laurence RG, Dor DM, Cohn LH, Bawendi MG, Frangioni JV (2004) Near-infrared fluorescent type II quantum dots for sentinel lymph node mapping. Nat Biotechnol 22:93–97. Epub 2003 Dec 2007.

Kirchner C, Liedl T, Kudera S, Pellegrino T, Munoz Javier A, Gaub HE, Stolzle S, Fertig N, Parak WJ (2005) Cytotoxicity of colloidal CdSe and CdSe/ZnS nanoparticles. Nano Lett 5:331–338.

Larson DR, Zipfel WR, Williams RM, Clark SW, Bruchez MP, Wise FW, Webb WW (2003) Water-soluble quantum dots for multiphoton fluorescence imaging in vivo. Science 300:1434–1436.

Lidke DS, Nagy P, Heintzmann R, Arndt-Jovin DJ, Post JN, Grecco HE, Jares-Erijman EA, Jovin TM (2004) Quantum dot ligands provide new insights into erbB/HER receptor-mediated signal transduction. Nat Biotechnol 22:198–203. Epub 2004 Jan 2004.

Michalet X, Pinaud F, Lacoste TD, Dahan M, Bruchez MP, Alivisatos AP, Weiss S (2001) Properties of fluorescent semiconductor nanocrystals and their application to biological labeling. Single Molecules 2:261–276.

Michalet X, Pinaud FF, Bentolila LA, Tsay JM, Doose S, Li JJ, Sundaresan G, Wu AM, Gambhir SS, Weiss S (2005) Quantum dots for live cells, in vivo imaging, and diagnostics. Science 307:538–544.

Miyawaki A, Sawano A, Kogure T (2003) Lighting up cells: labelling proteins with fluorophores. Nat Cell Biol Suppl:S1-7.

Mobbs P, Becker D, Williamson R, Bate M, Warner A (1994) Techniques for dye injection and cell labelling in Microelectrode techniques, The Plymouth workshop handbook., 2nd Edition. Edition. Cambridge, U.K.: Company of Biologists.
Morgan J, Huckfeldt R, Wong RO (2005) Imaging techniques in retinal research. Exp Eye Res 80:297–306.
Murray CB, Kagan CR, Bawendi MG (2000) Synthesis and characterization of monodisperse nanocrystals and close-packed nanocrystal assemblies. Annu Rev Mater Sci 30:545–610.
Niemeyer CM (2001) Nanoparticles, proteins, and nucleic acids: biotechnology meets materials science. Angewandte Chemie, International Edition 40:4128–4158.
Nirmal M, Brus L (1999) Luminescence Photophysics in Semiconductor Nanocrystals. Accounts of Chemical Research 32:407–414.
Olivier JC (2005) Drug transport to brain with targeted nanoparticles. NeuroRx 2:108–119.
Pan W, Kastin AJ (2004) Why study transport of peptides and proteins at the neurovascular interface. Brain Res Brain Res Rev 46:32–43.
Parak WJ, Gerion D, Pellegrino T, Zanchet D, Micheel C, Williams SC, Boudreau R, Le Gros MA, Larabell CA, Alivisatos AP (2003) Biological applications of colloidal nanocrystals. Nanotechnology 14:R15-R27.
Peng ZA, Peng X (2001) Formation of high-quality CdTe, CdSe, and CdS nanocrystals using CdO as precursor. Journal of the American Chemical Society 123:183–184.
Penn SG, He L, Natan MJ (2003) Nanoparticles for bioanalysis. Curr Opin Chem Biol 7: 609–615.
Pinaud F, King D, Moore HP, Weiss S (2004) Bioactivation and cell targeting of semiconductor CdSe/ZnS nanocrystals with phytochelatin-related peptides. J Am Chem Soc 126: 6115–6123.
Pinaud F, Michalet X, Bentolila LA, Tsay JM, Doose S, Li JJ, Iyer G, Weiss S (2006) Advances in fluorescence imaging with quantum dot bio-probes. Biomaterials 27:1679–1687. Epub 2005 Nov 1628.
Rieke F, Schwartz EA (1996) Asynchronous transmitter release: control of exocytosis and endocytosis at the salamander rod synapse. J Physiol 493 (Pt 1):1–8.
Roney C, Kulkarni P, Arora V, Antich P, Bonte F, Wu A, Mallikarjuana NN, Manohar S, Liang HF, Kulkarni AR, Sung HW, Sairam M, Aminabhavi TM (2005) Targeted nanoparticles for drug delivery through the blood-brain barrier for Alzheimer's disease. J Control Release 108:193–214. Epub 2005 Oct 2024.
Rosenthal SJ, Tomlinson I, Adkins EM, Schroeter S, Adams S, Swafford L, McBride J, Wang Y, DeFelice LJ, Blakely RD (2002) Targeting cell surface receptors with ligand-conjugated nanocrystals. J Am Chem Soc 124:4586–4594.
Salata OV (2004) Applications of nanoparticles in biology and medicine. Journal of Nanobiotechnology 2:No pp given.
Shiohara A, Hoshino A, Hanaki K, Suzuki K, Yamamoto K (2004) On the cyto-toxicity caused by quantum dots. Microbiol Immunol 48:669–675.
Silver J, Ou W (2005) Photoactivation of quantum dot fluorescence following endocytosis. Nano Lett 5:1445–1449.
Thomas P, Mortensen M, Hosie AM, Smart TG (2005) Dynamic mobility of functional GABAA receptors at inhibitory synapses. Nat Neurosci 8:889–897.
Tomlinson ID, Mason JN, Blakely RD, Rosenthal SJ (2005) Inhibitors of the serotonin transporter protein (SERT): the design and synthesis of biotinylated derivatives of 3-(1,2,3,6-tetrahydropyridin-4-yl)-1 H-indoles. High-affinity serotonergic ligands for conjugation with quantum dots. Bioorg Med Chem Lett 15:5307–5310. Epub 2005 Sep 5323.
Triller A, Choquet D (2003) Synaptic structure and diffusion dynamics of synaptic receptors. Biol Cell 95:465–476.
Vanmaekelbergh D, Liljeroth P (2005) Electron-conducting quantum dot solids: novel materials based on colloidal semiconductor nanocrystals. Chem Soc Rev 34:299–312.

Vu TQ, Maddipati R, Blute TA, Nehilla BJ, Nusblat L, Desai TA (2005) Peptide-Conjugated Quantum Dots Activate Neuronal Receptors and Initiate Downstream Signaling of Neurite Growth. Nano Letters 5:603–607.

West JL, Halas NJ (2003) Engineered nanomaterials for biophotonics applications: improving sensing, imaging, and therapeutics. Annu Rev Biomed Eng 5:285–292.

Wu X, Liu H, Liu J, Haley KN, Treadway JA, Larson JP, Ge N, Peale F, Bruchez MP (2003) Immunofluorescent labeling of cancer marker Her2 and other cellular targets with semiconductor quantum dots. Nat Biotechnol 21:41–46. Epub 2002 Dec 2002.

Yao J, Larson DR, Vishwasrao HD, Zipfel WR, Webb WW (2005) Blinking and nonradiant dark fraction of water-soluble quantum dots in aqueous solution. Proc Natl Acad Sci USA 102:14284–14289. Epub 12005 Sep 14216.

Zhai RG, Bellen HJ (2004) The architecture of the active zone in the presynaptic nerve terminal. Physiology (Bethesda) 19:262–270.

# Quantum Dot Methods for Cellular Neuroimaging

**Gabriel A. Silva**

## 1 Introduction

Nanotechnology and nanoengineering have the potential to produce significant scientific and technological advances in diverse fields, including biology and medicine. In a broad sense, they can be defined as the science and engineering involved in the design, syntheses, characterization, and application of materials and devices whose smallest functional organization in at least one dimension is on the nanometer scale, ranging from a few to several hundred nanometers. A nanometer is one billionth of a meter or three orders of magnitude smaller than a micron, roughly the size scale of a molecule itself (e.g., a DNA molecule is about 2.5-nm long while a sodium atom is about 0.2 nm). The potential impact of nanotechnology stems directly from the spatial and temporal scales being considered: materials and devices engineered at the nanometer scale imply controlled manipulation of individual constituent molecules and atoms in how they are arranged to form the bulk macroscopic substrate. This, in turn, results in nanoengineered substrates and devices that can be designed to exhibit specific and controlled bulk chemical and physical properties as a result of the control over their molecular synthesis and assembly.

This chapter is an adaptation of a review written by the author in *Progress in Brain Research* (Silva 2009), which collected a number of primary studies and published reviews by the author and colleagues into a single manuscript focusing on functionalized quantum dot imaging and specific labeling of neural cells and tissues, including both neurons and various types of glial cells (Pathak et al. 2006, 2007, 2009; Silva 2004, 2006). Specifically, this chapter discusses work done by the author's lab to optimize protocols for specific labeling and imaging of fixed,

G.A. Silva (✉)
Departments of Bioengineering and Ophthalmology and Neurosciences Program,
University of California, San Diego, CA, USA
e-mail: gsilva@ucsd.edu

G.A. Silva and V. Parpura (eds.), *Nanotechnology for Biology and Medicine: At the Building Block Level*, Fundamental Biomedical Technologies,
DOI 10.1007/978-0-387-31296-5_8, 

dissociated neural cells and cytoarchitecturally intact neural retinal tissue sections (Pathak et al. 2006, 2009), illustrates the use of such protocols and methods, and discusses the direct quantitative estimation of the number of functionally available antibodies conjugated to quantum dot nanocrystals for applications to biological labeling (Pathak et al. 2007). The interested reader is referred to other review articles by the author and colleagues that discuss applications of nanotechnologies to neuroscience (Provenzale and Silva 2009; Silva 2004, 2005, 2006, 2007, 2008a, b; Yu and Silva 2008).

## 2 Semiconductor Quantum Dots and Their Utility for Neurobiological Imaging

Semiconductor fluorescent quantum dots are nanometer-sized, functionalized particles that display unique physical properties that make them particularly well-suited for visualizing and tracking molecular processes in cells using standard fluorescence (Biju et al. 2008; Gao 2003; Jaiswal et al. 2004; Wu et al. 2003). They are readily excitable and have broad absorption spectra with very narrow emission spectra, allowing multiplexing of many different colored quantum dots; they display minimal photobleaching, thereby allowing molecular tracking over prolonged periods; they also display a blinking property that allows the identification of individual quantum dots. As a result, single-molecule binding events can be identified and tracked using optical fluorescence microscopy, allowing the pursuit of experiments that are difficult or not possible given other experimental approaches.

Quantum dots are nanometer-sized particles composed of a heavy metal core, such as cadmium selenium or cadmium telluride, with an intermediate unreactive zinc sulfide shell and a customized outer coating of different bioactive molecules tailored to a specific application (Fig. 1). The composition and very small size of quantum dots (5–8 nm) gives them unique and very stable fluorescent optical properties that are readily tunable by changing their physical composition or size. The photochemical properties of quantum dots allow selective fluorescent tagging of proteins similar to classical immunocytochemistry (ICC). However, the use of quantum dots is associated with minimal photobleaching and a much higher signal-to-noise ratio. Their broad absorption spectra but very narrow emission spectra allows multiplexing of many quantum dots of different colors in the same sample, something which cannot be achieved with traditional fluorophores. The physics responsible for these effects are beyond the scope of this brief introduction, but the small size of quantum dot particles results in large but specific energy jumps between the energy band gaps of excited electron–hole pairs in the semiconductor core. This effect results in scaled changes of absorption and emission wavelengths as a function of particle size so that small changes in the radius of quantum dots translate into very distinct changes in color. This physical property represents another major advantage over traditional organic fluorophores that, in general, require distinct chemistries to produce different colors. For biological applications, quantum dots

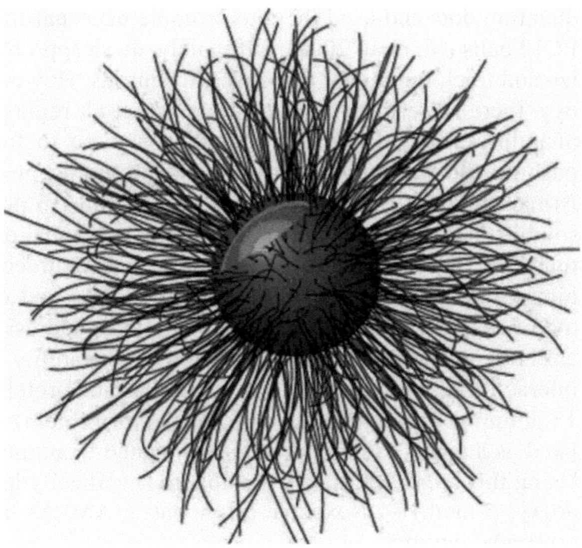

**Fig. 1** Cartoon structure of a typical functionalized quantum dot. The heavy metal core is shielded from the biological environment by an outer shell. The outer shell is in turn chemically functionalized with biologically relevant molecules, such as antibodies and other peptides (*black*), for specific binding to target epitopes, for example, on cells

can be chemically functionalized to target proteins at high ligand–receptor densities. Recent work has shown that, at least in some cellular systems, quantum dots conjugated with natural ligands are readily internalized into cells, do not interfere with intracellular signaling, and are nontoxic.

For neuroscience, quantum dots represent a tool of significant potential. Besides offering an alternative to traditional ICC, they are particularly valuable for studies of neurons and glia. Quantum dots can be used to visualize, measure, and track individual molecular events using fluorescence microscopy; they provide the ability to visualize and track dynamic molecular processes over extended periods (e.g., from seconds to many minutes). These properties are difficult to achieve using other techniques or approaches. For example, quantum dots are useful for experiments that are limited by the restricted anatomy of neuronal and glial interactions, such as the small size of the synaptic cleft, or between an astrocyte process and a neuron. Because of their extremely small size and optical resolution, they are also well-suited for tracking the molecular dynamics of intracellular and/or intercellular molecular processes over long timescales. However, it should be appreciated that the hydrodynamic radius of functionalized quantum dots is larger (15–20 nm) than their actual size of 5–8 nm. Recent studies using quantum dots in neuroscience illustrate the potential of this technology. Triller and colleagues used antibody functionalized quantum dots to track the lateral diffusion of glycine receptors in cultures of primary spinal cord neurons (Dahan et al. 2003). They were able to track the trajectory of individual glycine receptors for tens of minutes at spatial resolutions of 5–10 nm, demonstrating that the diffusion dynamics varied depending on whether the receptors were synaptic, perisynaptic, or extrasynaptic. Vu, Desai, and colleagues tagged nerve growth factor (bNGF) to quantum dots and used them to promote neuronal-like differentiation in cultured PC12 cells (Vu et al. 2005). Ultimately, these approaches could be used to visualize and track functional responses in neurons. However, as with any new technology, there are caveats. For example, Vu et al. reported that bNGF conjugated to quantum dots had reduced activity compared to free bNGF. Other groups are pushing the technology forward and providing new quantum dot-based tools. Brinker and colleagues developed a technique to produce biocompatible water-soluble quantum dot micelles that retain the optical properties of individual quantum dots. These micelles showed uptake and intracellular dispersion in cultured hippocampal neurons (Fan et al. 2005). Ting and colleagues are developing a modified quantum dot-labeling approach that addresses the relatively large size of antibody–quantum dot conjugates and the instability of some quantum dot–ligand interactions. Their technique tags cell surface proteins with a specific peptide (a 15 amino acid polypeptide called acceptor protein) that can be directly biotinylated as a target for streptavidin-conjugated quantum dots (Howarth et al. 2005). Using this approach, they were able to specifically label and track a-amino-3-hydroxyl-5-methyl-4-isoxazole-propionate (AMPA) receptors on cultured hippocampal neurons.

## 3 Specific Labeling and Imaging of Dissociated Neurons and Glial Cells

We have previously discussed in detail our quantum dot-labeling protocols for labeling neurons and glia (Pathak et al. 2006). We conjugated anti-b-tubulin III and anti-glial fibrillary acidic protein (GFAP) antibodies to 605-nm quantum dots and labeled primary cortical neurons, PC12 cells, primary cortical astrocytes, and r-MC1 retinal Muller glial cells. b-tubulin III and GFAP are ubiquitous cytoskeletal proteins specific to neurons and macroglia, respectively, but the protocols should label any protein of interest. Table 1 summarizes the detailed methods.

Using our protocols, we were able to get excellent specific labeling of b-tubulin in neurons and PC12 cells and GFAP in astrocytes and Muller cells, with negligible nonspecific binding or background (see Fig. 2). Labeling with unconjugated or primary antibody-omitted, streptavidin-conjugated quantum dots showed no labeling at all (data not shown). b-tubulin and GFAP labeling using functionalized quantum dots displayed similar labeling patterns to those expected using standard ICC controls visualized with fluorophore-tagged secondary antibodies (Fig. 2g, h). For comparable imaging conditions, quantum dot-labeled cells were brighter and displayed more detailed and sharper microstructural anatomy. The pattern of quantum dot labeling was typical for that observed in other cell types, displaying a dense punctuate pattern and fine details of both intracellular intermediate filaments and cellular processes, unlike traditional fluorophores which tend to have a diffused appearance due to the broad point spread function of their fluorescence signal. Nonspecific artifact labeling using some quantum dot protocols may label

**Table 1** Summary of quantum dot-labeling protocol for neurons and glia

| |
|---|
| Preprocessing and fixing |
| Remove media from wells by gently aspirating |
| Wash cells with warmed PBS |
| Fix cells with 4% paraformaldehyde (Electron Microscopy Sciences, catalog #157 15-S) in PBS for 10 min at room temperature |
| Wash cells 3× with PBS |
| Permeabilize cells with 0.2% Triton X-100 (Fisher Scientific, catalog #BP151-100) in PBS for 5 min |
| Wash cells 3× for 5 min with PBS |
| Incubate with 10% horse serum in PBS for 30 min at room temperature |
| Rinse with PBS |
| Apply Streptavidin/Biotin Blocking Kit (Vector Labs, catalog #SP-2002) |
| Primary incubation |
| Rinse with PBS |
| Add biotinylated molecule of interest (e.g., antibodies; use ProtOn Biotin Labeling Kit or similar for biotinylation; Vector Labs, catalog #PLK-1202) |
| Incubate for 2 h at room temperature (biotinylated secondary antibody for 1 h – alternative three-step-labeling protocol) |
| Remove antibodies by gentle aspiration and rinse 3× with PBS |
| Quantum dot incubation |
| Add streptavidin-conjugated quantum dots (we used Quantum Dot Corporation's 605-nm quantum dots here, catalog #1010-1) in 10% horse serum |
| Incubate for 1 h at room temperature |
| Rinse 3× with PBS |
| Mount with 90% glycerol (Sigma, catalog #G-6279) in PBS |

Reproduced from Pathak et al. (2006)

neural cells incorrectly due to nonspecific putative electrostatic interactions. We observed this when conjugating antibodies directly to quantum dots, which resulted in unconjugated quantum dots nonspecifically staining the nucleus of Muller cells (see Fig. 2i). Nonspecific binding was also observed when using other published protocols for nonneural cells (Wu et al. 2003). Blocking conditions also need to be carefully optimized since most standard blocking approaches did not work satisfactorily in our hands, including 1–5% bovine serum albumin, 10% horse serum, and 10% fetal bovine serum among others, which resulted in a high level of nonspecific quantum dot binding to the cells (data not shown). Another advantage to labeling with quantum dots is that each individually visualized dot in a fluorescence micrograph represents one to three individual quantum dots based on our own calculations and those of others (Chan and Nie 1998). This means that qualitative and potentially quantitative information can be measured for individual binding events between quantum dot-conjugated molecules and their cellular molecular targets, a direct result of the underlying physics (Michalet et al. 2005; West and Halas 2003) that cannot be done with standard ICC (see Dahan et al. 2003, for an example).

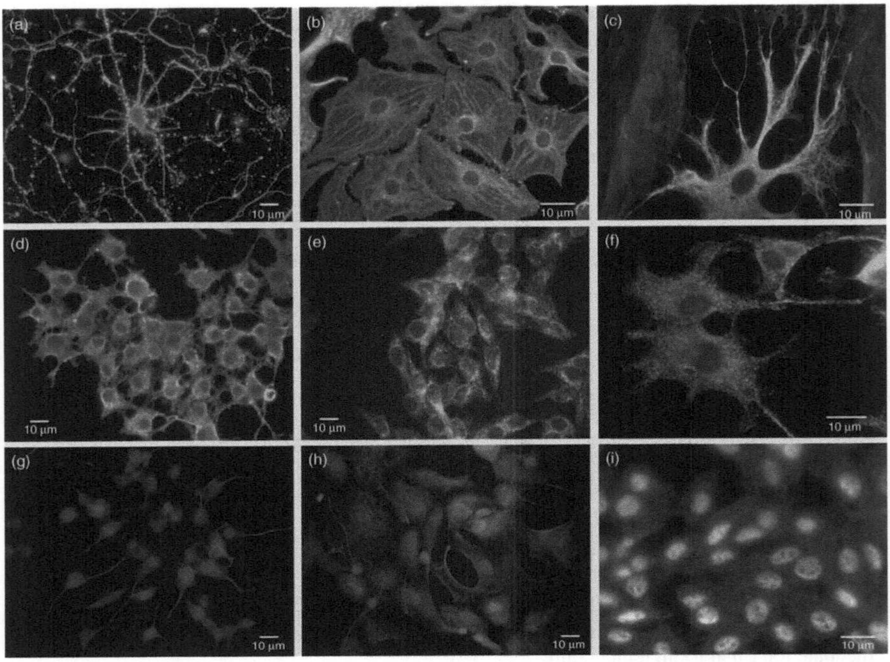


**Fig. 2** Fluorescent labeling of neurons and glia with antibody-conjugated 605-nm quantum dots. (**a**) Primary cortical neurons specifically labeled for b-tubulin. (**b, c**) Primary cortical astrocytes specifically labeled for GFAP. (**d, f**) PC12 cells labeled for b-tubulin. (**e**) r-MC1 neural retinal Muller glial cells specifically labeled for GFAP. (**g**) PC12 cells labeled for b-tubulin using standard ICC. (**h**) Primary spinal cord astrocytes labeled for GFAP using standard ICC. (**i**) An example of artifactual nonspecific labeling in r-MC1 Muller cells with anti-GFAP-conjugated 605-nm quantum dots. In this case, putative nonspecific electrostatic interactions between quantum dots and cellular proteins led to intense nuclear staining and mild cytoplasmic staining using other quantum dot conjugation protocols described for mammalian cells. All imaging parameters were constant for the different experimental conditions, with an acquisition/exposure time of 30 ms for all panels except (**i**), which was taken with an acquisition time of 100 ms. Reproduced from Pathak et al. (2006)

# 4 Efficacy of Different Antibody Conjugation Methods to Quantum Dots

One critical issue that has not been addressed is experimentally determining the number of antibodies bound to quantum dots which are functionally available for target protein binding (Pathak et al. 2007). This is critical for the analysis and proper interpretation of biological data labeled using these kinds of methods. While other groups have qualitatively characterized antibody-functionalized quantum dots using TEM, AFM, UV spectroscopy, and gel electrophoresis, and in some cases have suggested estimates of the putative number of total antibodies bound to quantum, no calculations of the number of functional antibodies bound to quantum dots based on quantitative experimental results have been reported. We previously reported derived numbers of functional IgG antibodies conjugated to quantum dots based on calculations of quantitative electrophoresis experiments using two different conjugation schemes: a common direct covalent conjugation using a reduced disulfide maleimide reaction and biotinylated antibodies bound to streptavidin-functionalized quantum dots. Antibody–quantum dot complexes were run in a sodium dodecyl sulfate-polyacrylamide gel electrophoresis (SDS-PAGE) to separate the functional component of conjugated antibodies from the quantum dots. We, then, blotted the antibody fragments onto a membrane to determine the identity and amount of the antibodies, and quantitatively compared the degree of functional binding to known protein standards to derive the number of bound antibodies.

The number of functional antibodies covalently bound to commercially available quantum dots was on average much less than one functional IgG molecule per quantum dot ($0.076 + 0.014$) and therefore of limited utility for biological experiments. In contrast, antibodies bound to quantum dots via the streptavidin–biotin system resulted in higher numbers of functional antibodies, with $0.60 + 0.14$ IgG molecules per quantum dot for a 1:1 IgG:quantum dot molar ratio and $1.3 + 0.35$ IgG molecules per quantum dot for a 2:1 ratio. In addition to these specific results, our methods may be of broader interest because our approach is easily extendable for experimentally deriving the number of functional antibodies or peptides bound to other classes of nanoparticles (e.g., magnetic nanoparticles).

We begin by considering the covalent conjugation of antibodies to quantum dots. Prior to their conjugation, antibodies were reduced using dithiothreitol (DTT), which generates three distinct fragments identifiable by their molecular weights: a 25-kDa light chain, which importantly includes the functional-specific epitope-binding site for a particular IgG molecule; a 50-kDa heavy chain; and a 75-kDa partially cleaved chain consisting of a heavy chain and a light chain held together by an unreduced disulfide bond (Fig. 3a). Following this, individual fragments were covalently bound to quantum dots via an *N*-succinimidyl 4-(maleimidomethyl) cyclohexanecarboxylate (SMCC) linkage bond which cannot be broken by DTT treatment, an important consideration for the interpretation of the experimental results that follow. This gives rise to three possible binding scenarios to quantum dots (Fig. 3a): covalently bound light chains, covalently bound heavy chains, and covalently bound heavy–light chain partial fragments, of which only the latter can undergo further DTT reduction to remove the light chain fragment from heavy chains that remain bound to quantum dots or heavy chains removed from light chains bound to quantum dots.

We first confirmed that antibodies were indeed covalently bound to the quantum dots by running IgG–quantum dot complexes though SDS-PAGE with and without DTT. For DTT-reduced conditions, we observed light chains cleaved from covalently bound partial fragments (Fig. 4a, lanes 4–6). As expected, this separation occurred minimally in lanes without DTT (Fig. 4a, lanes 2 and 3). The presence of a weak band at the 25-kDa position in nonreduced lanes (Fig. 4a, lanes 2 and 3) was due to low concentrations of reducing agents in the gel and running buffers. Interestingly, we saw no heavy chains being dissociated from light chain-bound

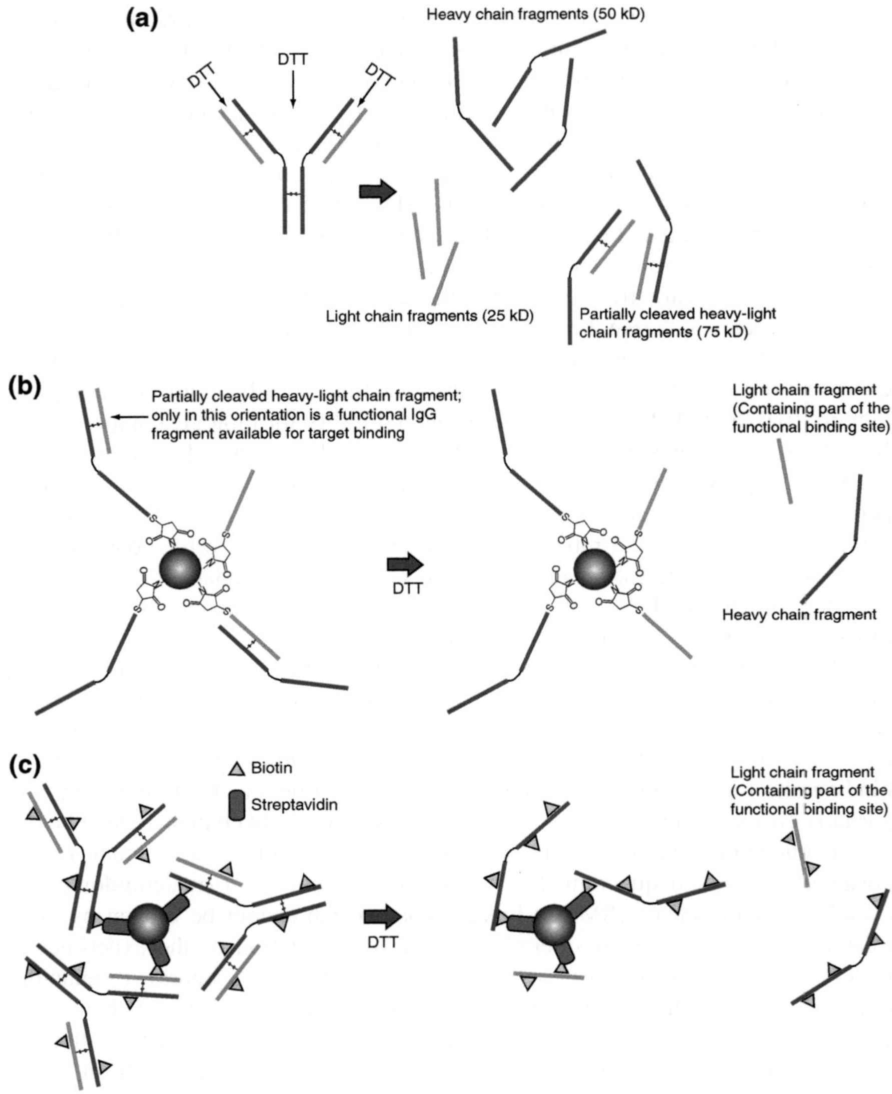


**Fig. 3** Antibody reduction and conjugation to quantum dots. (**a**) Schematic of antibody cleavage sites by DTT at disulfide linkages. The fragments that can result from DTT reduction include the light chain, heavy chain, and partially cleaved fragments due to incomplete reduction. (**b**) Schematic of direct SMCC covalent conjugation of antibodies to quantum dots. Further reduction with DTT following the primary reduction associated with the conjugation reaction yields the light chains which are counted in the derivation of the average number of functional IgG molecules originally on quantum dots. (**c**) Similar schematic for biotinylated antibodies conjugated to streptavidin-coated quantum dots. Reproduced from Pathak et al. (2007)

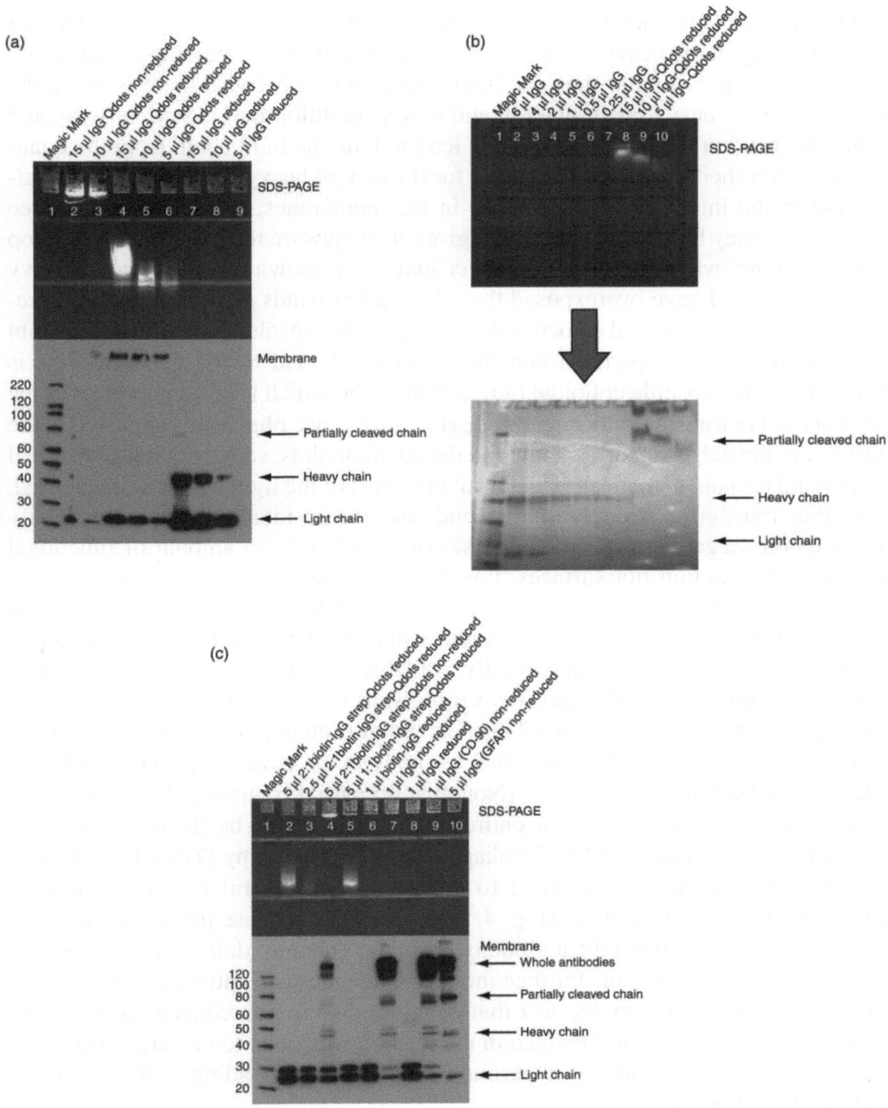


**Fig. 4** Separation of IgG antibodies into fragments using SDS-PAGE and membrane transfer under different experimental and control conditions. (**a**) Covalently conjugated IgG to quantum dots via an SMCC linker and controls. (**b**) Colloidal nonspecific stain for proteins in gels for the direct conjugation method. (**c**) Biotinylated IgG bound to streptavidin-coated quantum dots and controls. Reproduced from Pathak et al. (2007)

partial fragments. It is unclear why this was the case, although we hypothesize that the probability of the heavy chain portion of a partial fragment binding to a quantum dot is considerably higher than the light chain portion because there is twice the surface area for heavy chain binding and it is a condition that is sterically favored (since the bend in the partial chain may tend to hide the light chain from the quantum dot). Another potential explanation for the lack of heavy chain is methodological. Given the intensity of other bands in the membranes, small amounts of free heavy chain may have gone undetected given the exposure time we used to develop the membrane, which if had been longer may have shown the presence of heavy chains but would have overexposed the other darker bands resulting in uninterpretable smearing. Additional evidence that heavy chains covalently bound to quantum dots originating from partial fragments remained bound to the quantum dots is inferred by a nonspecific colloidal blue protein stain, which labels any protein in the gel that did not transfer to the membrane (Fig. 4b). Since blue bands appeared at the position in the gels that corresponded to the quantum dots, some amount of residual protein did remain on their surface. Given that most of the light chains were cleaved, since they transferred strongly to the membrane, this residual protein is most likely heavy chain. Regardless, for the purposes of calculating the amount of functional antibody on quantum dot surfaces, this is of minimal importance, since it is the amount of available light chain that we are interested in since it is the light chain that contains the ligand-binding epitope. Another important consideration to note is that the amount of partial fragments initially available for binding to quantum dots following the initial DTT reduction was very low, as evident in the reduced unconjugated IgG controls (Fig. 4a, lanes 7–9). This point is an important consideration for why the number of available functional antibody in the covalently conjugated condition was calculated to be so low. (Note that no partial fragments were visible for the quantum dot lanes because the entire partial chain cannot be cleaved intact from the quantum dot since the SMCC linkage cannot be broken by DTT.) If antibodies had been electrostatically attached to quantum dots, several bands would have shown up in nonreduced lanes (Fig. 4, lanes 2 and 3) because the gel would have electrostatically separated the antibodies from the quantum dots according to their molecular size and weight. Further indirect evidence that antibodies were covalently bound is implied by the fact that quantum dots in nonreduced lanes did not travel through the gels but remained in the loading wells due to the large size of the unreduced complex (visible as hyperintense signals in the loading wells for lanes 2 and 3 of the SDS-PAGE).

We ran the same experiments with biotinylated antibodies and streptavidin-coated quantum dots at 2:1 and 1:1 antibody-to-quantum dot molar ratios. Biotinylated antibodies have biotin molecules throughout the entire antibody, which results in the IgG molecules being conjugated to quantum dots presumably in all possible spatial arrangements (Fig. 4c). Importantly and very differently from the direct covalent conjugation reaction, using the biotin–streptavidin system, the entire antibody molecule is conjugated to the quantum dot; it is not reduced into its light chain and heavy chain fragments prior to binding. Similar to the covalent antibody conjugation method, nonreduced conditions resulted in quantum dots remaining in the loading wells (Fig. 4c, lane 4) while reduced conditions allowed quantum dots to run through the gels (Fig. 4c, lanes 2, 3, and 5). Some amount of antibodies did transfer in nonreduced conditions for biotin–streptavidin, IgG–quantum dot complexes because the reducing agents in the running buffers and the gel caused the light chain to dissociate in the same manner as for the covalent conjugation. However, since all bands were much stronger in the biotin–streptavidin method in general, bands for the non-reduced condition were correspondingly stronger. Bands in non-DTT-treated antibody lanes (i.e., Fig. 4c, lanes 7, 9, and 10) show the reduction process in greater detail since reduction agents in the running buffers reduced the antibodies less efficiently than DTT-treated conditions (Fig. 4a, lanes 2, 3, 5, 6, and 8).

Based on these data and the qualitative models introduced above that describe the different putative binding scenarios for antibodies directly covalently conjugated to quantum dots and for antibodies bound to quantum dots via biotin and streptavidin (Fig. 3b), we derived the average number of functional IgG conjugated to quantum dots. We use the term "functional antibody" to describe the amount of Fc light chain, which includes a part of the target protein-binding epitope that is physically oriented outward from a quantum dot and presumably able to interact with its ligand. Molecularly, roughly, the first 110 amino acids at the amino-terminal end of both heavy and light chains form the variable V regions which contain highly variable segments called complementary-determining regions. The pairwise association of V regions from both heavy and light chains is what actually forms the antigen-binding site. As such, only a partial fragment bound to the quantum dot would be functional. Furthermore, because of the structure of the antigen-binding site, a partial fragment covalently bound to the quantum dot oriented with the light chain facing the nanoparticle would almost surely prevent ligand binding. Since it is the Fc light chain portion of the antibody that actively binds to proteins, quantifying the amount of light chain fragments not directly bound to quantum dots and oriented outward gives a good approximation of the functional activity of antibody–quantum dot complexes.

To determine the number of functional IgG bound to quantum dots, we measured the density of the 25-kDa light chain bands and compared them to controls of known antibody concentrations. Using image analysis software that measures the band density of electrophoresis gels (ImageQuant TL, GE Healthcare), we fitted curves to known concentrations of unconjugated IgG to obtain standard curves of IgG band densities (Fig. 4a and b; all $r^2 \geq 0.89$). Using these curves, we then determined the concentration of IgG bands associated with covalently bound IgG and 2:1 and 1:1 IgG:quantum dot molar ratio streptavidin–biotin conjugation conditions (Fig. 5c, d). Finally, we calculated the number of functional antibodies bound to the quantum dots for each condition (Fig. 5e).

For covalently conjugated IgG, we calculated that on average there is much less than one antibody molecule (0.076 + 0.014) per quantum dot. In other words, adding 10 mL of antibodies directly conjugated to quantum dots is equivalent to adding 0.455 mL from a 0.5 mg/mL stock. This suggests that covalently conjugated antibodies have low amounts of functionally available antibodies and are of inadequate

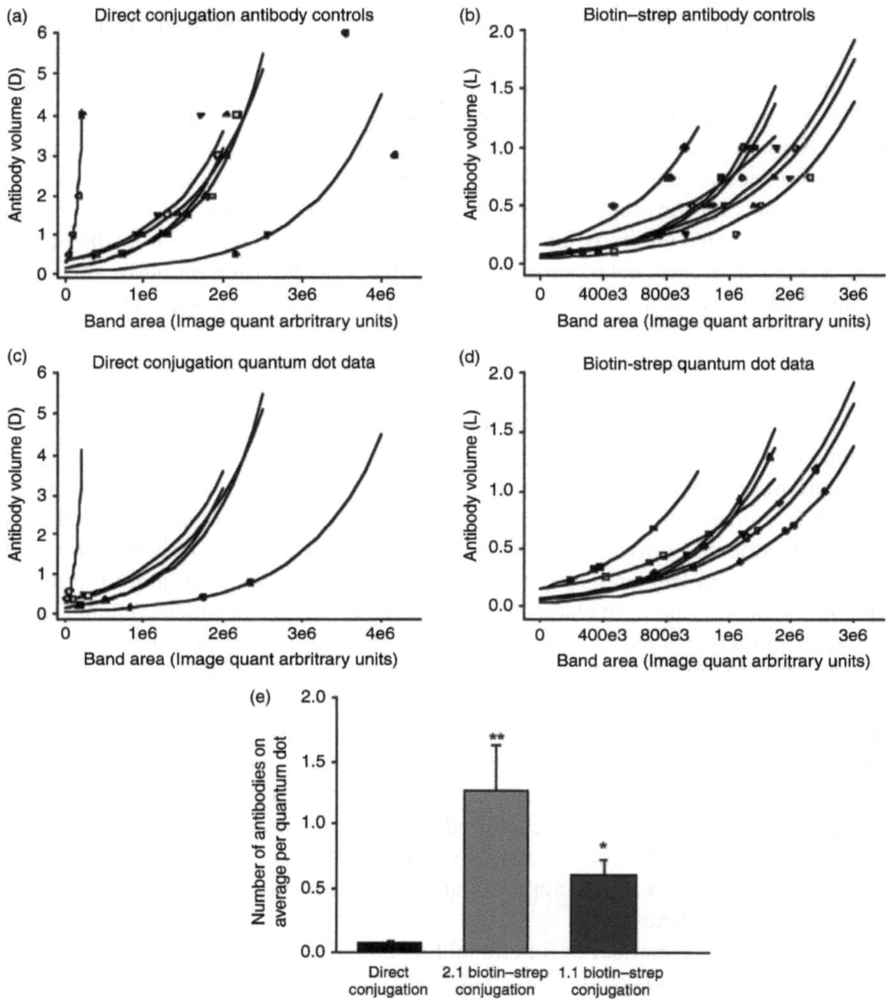


**Fig. 5** Derivation of the average number of functional antibodies on both covalently conjugated and streptavidin–biotin-conjugated quantum dots based on measurements of the bound density for different concentrations. (**a**, **b**) Fitted linear log control curves (ln $y = ax \bullet b$) for known volumes of unconjugated IgG antibody band densities in SDS-PAGE gels. Note that the data for each gel was fitted with its own curve in order to control for intergel variability. Each symbol represents a different gel ($n = 6$ gels for covalently conjugated IgG conditions containing a total of 32 unconjugated IgG controls and 13 IgG–quantum dot complexes, and 7 gels for streptavidin–biotin, IgG–quantum dot complexes containing a total of 35 unconjugated IgG controls and 28 IgG–quantum dot complexes). (**c**, **d**) Corresponding derived volumes from SDS-PAGE band densities for conjugated and streptavidin–biotin-conjugated antibody–quantum dot complexes using the curves plotted in panels (**a**) and (**b**), respectively. (**e**) Calculated values for the average number of antibodies conjugated to quantum dots for both conditions based on the derived measurements of functional antibody volumes (* and ** $p < 0.01$). Reproduced from Pathak et al. (2007)

sensitivity for reliable specific labeling of target proteins. In contrast, the number of antibodies bound to quantum dots via the strepavidin–biotin system resulted in a more biologically reasonable 0.60 + 0.14 IgG molecules per quantum dot for a 1:1 IgG:quantum dot molar ratio and, as would be expected, 1.3 + 0.35 IgG molecules per quantum dot for a 2:1 ratio. This is equivalent to a functional volume of 0.943 mL of antibody for a 2:1 molar ratio or 0.53 mL for 1:1 molar ratio starting from 4 mL of biotinylated antibody conjugated to streptavidin quantum dots from a 0.5 mg/mL stock concentration. We acknowledge that these numbers are an approximation, since light chains near the quantum dot surface attached to a heavy chain bound to the quantum dot as part of a partial fragment would be sterically unavailable for antigen binding but could still dissociate following DTT reduction. However, this may represent a small source of error because it may be sterically difficult for bound heavy–light chain domains to bind to the quantum dot, therefore thermodynamically favoring the functional partial fragment orientation (see Fig. 3b). In any case, this error would contribute to an overestimation of the number of functional antibodies conjugated to a quantum dot, and therefore represent an upper bound on the number of putative functional antibodies, further emphasizing the significance of the results we present here.

These results are significantly less than the suggested estimates of about two to ten antibodies conjugated per quantum dot. To the best of our knowledge, no conjugation reaction can control the binding orientation of IgG molecules. Consequently, due to Brownian motion, the number of bound functional antibodies is almost certainly less than the number of total bound IgG. This is not considered by TEM imaging approaches that measure the size (i.e., diameter) of antibody–quantum dot complexes in order to estimate the number of bound antibodies. An important question is: Why did covalent conjugations result in lower numbers of functional antibodies compared to streptavidin–biotin conjugations? One possible explanation is that DTT-reduced antibody fragments attaching to the surface of quantum dots leave few opportunities for light chain fragments to be properly oriented outward and available for protein binding, since of the three reduced fragment types only partial fragments result in functional antibodies and even then the orientation of the partial fragment binding to the quantum dot surface must be correct to allow the light chain fragment to point outward in order to interact with its ligand. In biotin–streptavidin conjugations, the antibody is never cleaved, leaving the whole molecule bound to the quantum dot surface and structurally offering more opportunities for light chain fragments to bind their targets. It is plausible that other covalent, conjugation chemistries result in higher yields of functional antibodies, comparable to those we report for streptavidin–biotin conjugates or even higher, but it cannot simply be assumed so since, as we show here, at least one well-established and commonly used covalent conjugation reaction results in very low numbers of functional antibody on quantum dots. We propose that functionalized quantum dot labeling of biological preparations need to be preceded by the experimental determination of the number of functionalized antibodies per quantum dot, especially given the variability in conjugation methods between different labs. These considerations have a direct impact on the quality, interpretation, and relevance of biological or physiological results obtained using quantum dot-labeling nanotechnologies.

## 5 Labeling Reactive Gliosis in Retinal Tissue Sections

In the normal neural retina, GFAP expression is associated with the astrocyte layer in the inner nuclear layer and the endfeet of Müller glial cells near the retinal capillaries. Quantum dot labeling of GFAP in control sections of rat retina showed that only Müller cell endfeet and astrocytes were GFAP positive, with no GFAP upregulation and no nonspecific binding (Fig. 6a, b). The high specificity and signal-to-noise ratio

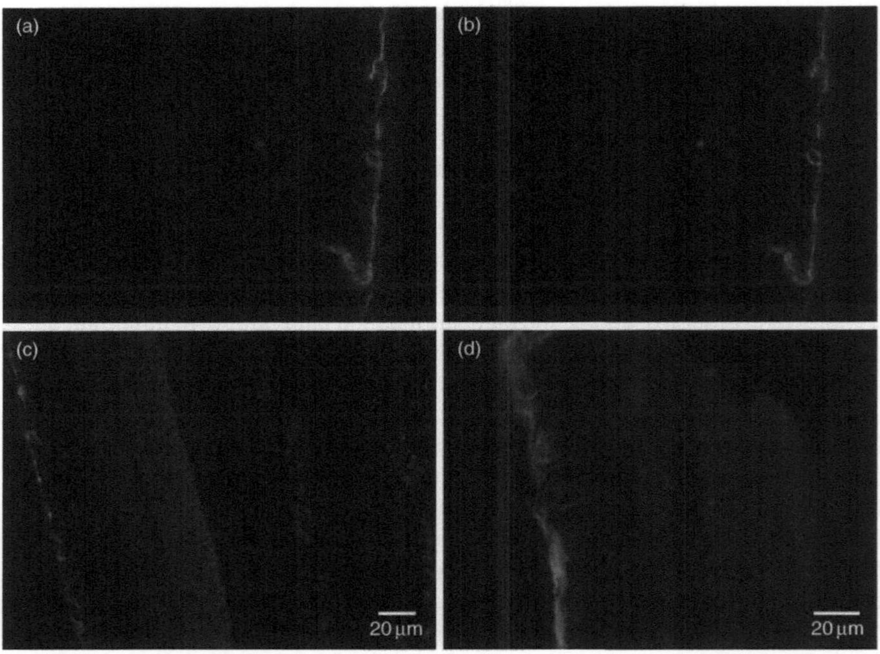


**Fig. 6** Control labeling of the noninjured rat neural sensory retina for GFAP. (**a, b**) Anti-GFAP antibody functionalized quantum dot conjugates specifically label only Muller cell endfeet-associated retinal capillaries and astrocytes associated in the inner nuclear layer associated with retinal ganglion cells. Two slices from a wide-field nonconfocal image stack are shown, and display no observable nonspecific labeling despite the use of nonconfocal mode. (**c, d**) Wide-field nonconfocal standard ICC using an anti-40,6-diamidino-2-phenylindole(GFAP)-conjugated primary antibody and FITC fluorophore-tagged secondary antibody. A nonspecific nuclear DAPI stain was used to visualize the other retinal layers. Note the more diffuse labeling using FITC compared to the quantum dots and the presence of some nonspecific labeling in the distal layers of the retina. Panels (**a**), (**b**), and (**d**) were taken at 40× and 50-ms exposure times while panel (**c**) was taken at 20× at a 50-ms exposure. All micrographs are 10-mm slices. Reproduced from Pathak et al. (2009)

of our quantum dot-labeling protocol are particularly emphasized in Fig. 6 because the micrographs were taken in wide-field nonconfocal mode which collects light from the entire thickness of the tissue slice, unlike in confocal mode, where stray light is physically excluded from the plane of focus. This is critical because, as discussed above, one of the biggest difficulties associated with immunospecific quantum dot labeling of neural cells is nonspecific interactions and clumping between quantum dot particles that can produce false-positive results (Pathak et al. 2006, 2007). Methodologically, nonspecific binding proved to be more of an issue with the methanol-fixed samples; all samples shown in the results were fixed using paraformaldehyde followed by Triton X-100 to remove excessive cross-linking of proteins induced by fixation. It is essential for the bulkier quantum dot conjugates, compared to the smaller size of traditional fluorophores, to experience a low amount of cross-linking in the tissue in order to avoid clumping. It is a common mistake to assume that quantum dot nanoparticles are smaller than fluorescent dyes, when in fact they are 10–20 times larger (depending on the color) than fluorescein isothiocyanate (FITC). Serially sectioned 10-mm slices throughout the thickness of the retina, of which two consecutive slices are shown in Fig. 6a, b, showed no observable nonspecific labeling. These results are similar to control retinal sections labeled using traditional ICC with a primary antibody specific to the target antigen and an FITC-tagged secondary antibody that binds to the primary antibody and acts as a fluorescent reporter (Fig. 7c, d), although the FITC labeling was somewhat more qualitatively diffuse and did show some degree of nonspecific labeling despite our best attempts.

Upregulation of GFAP in Müller cells and astrocytes occurs only under pathological conditions and is considered the hallmark of the reactive glial response. GFAP upregulation in Müller cells is particularly apparent because it spans the length of their cell bodies throughout most of the thickness of the retina up to the inner limiting membrane. In the rat laser-induced choroidal neovascularization (CNV) model we used, gliosis and glial scarring occur as secondary processes and result in a strong upregulation of GFAP. Figure 8 shows a confocal z-stack of a 10-mm tissue section with an imaged slice thickness of 1 mm centered at a laser-induced lesion site. To the best of our knowledge, these results represent the first successful specific labeling in situ of an intact neural tissue preparation. The intense upregulation of GFAP in both Müller cells and astrocytes indicated a strong reactive response to the induced trauma. Our quantum dot-labeling protocol was optimized to ensure even tissue penetration, minimal nonspecific antigen labeling, and maximal specific antigen retrieval. Given this, the fact that the upregulation in GFAP for all lesions we looked at extended over a cross-sectional thickness of the retina of about 10 mm, as demonstrated in Fig. 8 by the drop in fluorescence signal in the confocal stack by slice 10, suggests that the reactive volume of the neural retina in response to the laser-induced injury averaged between 9 and 10 mm in cross-sectional width. The high signal-to-noise ratio of the quantum dot-labeling procedure also putatively provides greater observable and therefore measurable cellular detail throughout the volume of the glial response. In the representative stack in Fig. 8, the upregulation progresses from proximal Müller cell processes near the boundary of the lesion site (progressively from slice 1 to 3) to the entire length of the Müller cells and astrocyte layer near the center of the lesion

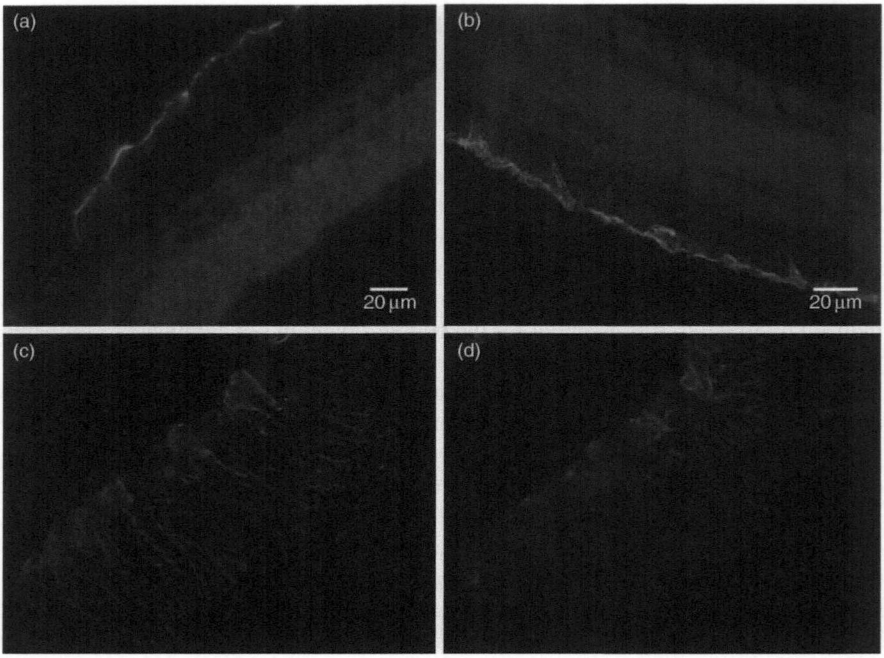


**Fig. 7** Specific labeling of GFAP upregulation in the rat neural retina at a laser-induced lesion site imaged using standard FITC ICC. (**a, b**) Wide-field nonconfocal standard ICC using an anti-GFAP-conjugated primary antibody and FITC fluorophore-tagged secondary antibody. A nonspecific nuclear DAPI stain was used to visualize the other retinal layers. (**c, d**) Confocal imaging of two different lesions with a 1.6-mm optical slice and an acquisition time of 2 s, comparable with the data shown in Fig. 2. (**c**) Shows a slice near the center of a lesion while (**d**) shows a slice closer to the boundary of a lesion. Note the particularly high background and diffuse labeling in (**d**). Reproduced from Pathak et al. (2009)

(slices 4–6), followed by a progressive visible decrease in reactivity near the other side of the lesion boundary (progressively from slice 7 to 9). Therefore, this labeling method should be amenable to quantitatively measuring the extent and thickness of glial scars and presumably other neuronal- and glial-specific markers in neural tissue preparations at high spatial resolutions due to the cellular specificity and low background of the procedure. Such an approach would conceivably allow better quantitative measurements and statistics of both physiologically normal and, as illustrated here, pathological cellular processes. This quantum dot-labeling procedure is considerably superior to nonconfocal wide-field epifluorescence microscopy of retinal sections for specific labeling and imaging of GFAP upregulation in gliosis due to diffuse labeling and higher nonspecific background in the latter (Fig. 7a, b), as introduced above. In our hands, the quantum dot-labeling procedure was subjectively less diffuse, more intense, had a noticeably lower nonspecific background, and showed more cellular detail than the best-optimized standard FITC ICC labeling we could achieve (Fig. 7c, d). This was especially true near the border of imaged lesions, where GFAP

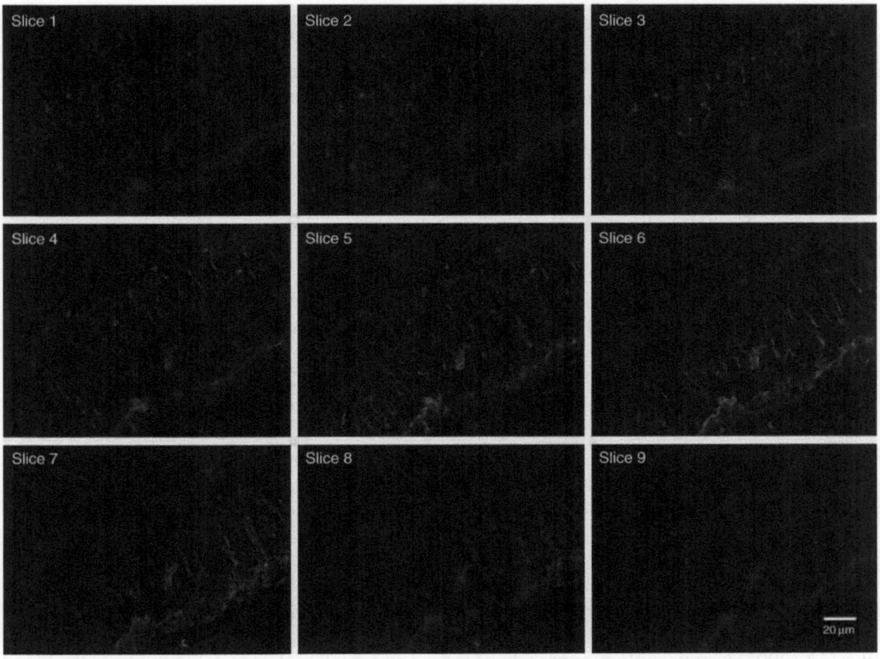


**Fig. 8** Specific labeling of GFAP upregulation in the rat neural retina at a laser-induced lesion site imaged using anti-GFAP quantum dot conjugates. A serial cross section of the retina 10-mm thick that encompassed one of the induced lesions was imaged at 1-mm thick optical slices using confocal microscopy using a 1.8-s exposure time. Reproduced from Pathak et al. (2009)

upregulation gradually decreased and there was less signal intensity (Fig. 7d). In these cases, FITC labeling appeared to display considerably higher nonspecific background, which makes this approach difficult for identifying the edges of GFAP upregulation and gliosis lesion boundaries, resulting in less confidence in any derived measurements of lesion volumes and the spread of the reactive glial response.

## References

Biju V, Itoh T, Anas A, Sujith A, Ishikawa M (2008) Semiconductor quantum dots and metal nanoparticles: Syntheses, optical properties, and biological applications. Anal Biochem 391:2469–2495.

Chan WC, Nie S (1998) Quantum dot bioconjugates for ultrasensitive nonisotopic detection. Science 281:2016–2018.

Dahan M, Levi S, Luccardini C, Rostaing P, Riveau B, Triller A (2003) Diffusion dynamics of glycine receptors revealed by single-quantum dot tracking. Science 302:302, 442–445.

Fan H, Leve EW, Scullin C, Gabaldon J, Tallant D, Bunge S (2005) Surfactant-assisted synthesis of water-soluble and biocompatible semiconductor quantum dot micelles. Nano Letters 5:645–648.

Gao X (2003) Molecular profiling of single cells and tissue specimens with quantum dots. Trends Biotechnol 21:371–373.
Howarth M, Takao K, Hayashi Y, Ting A (2005) Targeting quantum dots to surface proteins in living cells with biotin ligase. Proceedings of the National Academy of Sciences 102:7583–7588.
Jaiswal JK, Goldman ER, Mattoussi H, Simon SM (2004) Use of quantum dots for live cell imaging. Nature Methods 1:73–78.
Michalet X, Pinaud FF, Bentolila LA, Tsay JM, Doose S, Li JJ, Sundaresan G, Wu AM, Gambhir SS, Weiss S (2005) Quantum dots for live cells, in vivo imaging, and diagnostics. Science 307:538–544.
Pathak S, Davidson MC, Silva GA (2007) Characterization of the functional binding properties of antibody conjugated quantum dots. Nano Letters 7:1839–1845.
Pathak S, Cao E, Davidson MC, Jin S, Silva GA (2006) Quantum dot applications to neuroscience: New tools for probing neurons and glia. Journal of Neuroscience 26:1893–1895.
Pathak S, Tolentino R, Kim N, D'Amico L, Barron E, Cheng L (2009) Quantum dot labeling and imaging of glial fibrillary acidic protein intermediate filaments and gliosis in the rat neural retina and dissociated astrocytes. Journal of Nanoscience and Nanotechnology 9:5047–5054.
Provenzale JM, Silva GA (2009) Applications of nanoparticles to central nervous system imaging and therapy. American Journal of Neuroradiology 10:3174.
Silva G (2004) Introduction to nanotechnology and its applications to medicine. Surgicial Neurology 61:216–220.
Silva GA (2005) Nanotechnology approaches for the regeneration and neuroprotection of the central nervous system. Surgicial Neurology 63:301–306.
Silva GA (2006) Neuroscience nanotechnology: Progress, opportunities and challenges. Nature Reviews 7:65–74.
Silva GA (2007) Nanotechnlogy approaches for drug and small molecule delivery across the blood brain barrier. Surgicial Neurology 67:113–116.
Silva GA (2008a) The Central Nervous System. Drug Discovery Today: Disease Models 5:1–3.
Silva GA (2008b) Nanotechnology approaches to crossing the blood-brain barrier and drug delivery to the CNS. BMC Neuroscience 9:S4.
Silva GA (2009) Quantum dot nanotechnologies for neuroimaging. Progress in Brain Research 180:19–34.
Vu TQ, Maddipati R, Blute TA, Nehilla BJ, Nusblat L, Desai TA (2005) Peptide-conjugated quantum dots activate neuronal receptors and initiate downstream signaling of neurite growth. Nano Letters 5:603–607.
West JL, Halas NJ (2003) Engineered nanomaterials for biophotonics applications: improving sensing, imaging, and therapeutics. Ann Biomed Eng 5:285–292.
Wu X, Liu H, Liu J, Haley K, Treadway J, Larson J, Ge N, F. P, Bruchez MP (2003) Immunofluorescent labeling of cancer marker Her2 and other cellular targets with semiconductor quantum dots. Nat Biotechnol 21:41–46.
Yu D, Silva GA (2008) Stem cell sources and therapeutic approaches for central nervous system and neural retinal disorders. Neurosurgical Focus 24:E11.

# Carbon Nanotubes as Electrical Interfaces to Neurons

**Michele Giugliano, Luca Gambazzi, Laura Ballerini, Maurizio Prato, and Stephane Campidelli**

## Abbreviations

| | |
|---|---|
| CNS | Central nervous system |
| $C_m$ | Membrane capacitance |
| $C_{cnt}$ | CNT equivalent capacitance |
| CNT | Carbon nanotube |
| $C_{sh}$ | Amplifier shunt capacitance |
| $E_K$ | $K^+$ Nerst equilibrium potential |
| $E_L$ | Nerst equilibrium potential of leak membrane currents |
| $E_{Na}$ | $Na^+$ Nerst equilibrium potential |
| FET | Field-effect transistor |

M. Giugliano
Laboratory of Neural Microcircuitry, Brain Mind Institute, École Polytechnique Fédérale de Lausanne, Lausanne, Switzerland

Department of Biomedical Sciences, University of Antwerp, Belgium

L. Gambazzi
Laboratory of Neural Microcircuitry, Brain Mind Institute, École Polytechnique Fédérale de Lausanne, Lausanne, Switzerland

L. Ballerini (✉)
Life Science Department, B.R.A.I.N., University of Trieste, Trieste, Italy
e-mail: lballerini@units.it

M. Prato
Department of Pharmaceutical Sciences, University of Trieste, Trieste, Italy

S. Campidelli
Laboratoire d'Electronique Moléculaire, CEA Saclay, Gif-sur-Yvette, France

G.A. Silva and V. Parpura (eds.), *Nanotechnology for Biology and Medicine: At the Building Block Level*, Fundamental Biomedical Technologies,
DOI 10.1007/978-0-387-31296-5_9, © Springer Science+Business Media, LLC 2012

| | |
|---|---|
| $g_K$ | $K^+$ conductance |
| $g_L$ | Leak conductance |
| $g_{Na}$ | $Na^+$ conductance |
| MWNT | Multi-walled carbon nanotube |
| $R_a$ | Axial cytoplasmic resistance |
| $R_{cnt}$ | CNT equivalent resistance |
| $R_{in}$ | Amplifier input resistance |
| $R_s$ | Seal resistance |
| $R_{sp}$ | Spread resistance |
| SWNT | Single-walled carbon nanotube |
| $V_{cnt}$ | CNT potential |
| $V_m$ | Membrane potential |

# 1 Introduction

Translating basic neuroscience research into experimental neurology applications often requires functional interfacing of the central nervous system (CNS) with artificial devices designed to monitor and/or stimulate brain electrical activity. Ideally, such interfaces should provide a high temporal and spatial resolution over a large area of tissue during stimulation and/or recording of neuronal activity, with the ultimate goal to elicit/detect the electrical excitation at the single-cell level and to observe the emerging spatiotemporal correlations within a given functional area. Activity patterns generated by CNS neurons have been typically correlated with a sensory stimulus, a motor response, or a potentially cognitive process.

The growing interest in interfacing CNS structures to artificial devices is related to the possible improvements in our ability to decode and interpret brain signals. This achievement might be of critical importance not only for elucidating how the brain works, but also to reach a control of prosthetic devices by pure thought. The progress in neuronal activity recording and/or stimulation strongly relies on the optimization of neuron–electrode functional interface, transducing ion fluxes related to neuronal electrical activity into electron fluxes within the electrode-conducing material. Such a functional coupling has been traditionally achieved by means of metal electrodes, which allow variable degrees of neuronal target selectivity, signal sensitivity, and recording distortions. The ongoing amelioration of these important features is addressed by optimizing the mechanical and physical properties of the electrodes.

In basic neuroscience research, the application of neuronal recordings contributed a great deal to the global comprehension of CNS operation, and each technical achievement or improvement has been rewarded by significant scientific discoveries. For instance, advances in recording technologies, such as the development of the glass micropipette, have provided researchers with excellent and cheap transduction interfaces whose size matches that of subcellular structures. This has allowed the quantitative study of the ionic basis of nerve transmission. Thus, although the evolution of electrode/CNS interfaces is likely to remain driven by important clinical goals, it also offers unique tools for challenging the main ideas of modern neuroscience.

The issue of the interface between the electrode and the neuronal tissue has been addressed by biomaterials and tissue-engineering approaches. It has been shown that material texture or topographic features are crucial in promoting attachment and growth of neurites in vitro, significantly implementing stability of recordings (Schwartz et al. 2006). Indeed, efforts to improve implant design promote the development of more reliable and long-term recordings. The observation that engineered nanoscale features, obtained by layering nanotubes on a substrate surface, encourage neuronal growth while inhibiting astrocyte attachment (Moxon et al. 2004) hints at the idea of exploiting the technological aspects of nanomaterials to optimize interfaces between a technical device and the neuronal environment.

Current developments of nanotechnology applications to biology and medicine are strongly based on the capacity of nanomaterials to interact with biological systems at a subcellular level. The ability to chemically control nanosubstrates produces materials, which functionally integrate with cellular and physiological systems to an extraordinary degree (Giugliano et al. 2008). In particular, this latter interaction can be understood and engineered with a high level of specificity (Silva 2006). The field of neuronal interfaces is particularly related to such an interdisciplinary approach, and the mastering of new conductive nanoparticles is paving the way to a novel kind of neuroelectronics. Nanotechnology applications to the CNS have the potential to provide a new paradigm for the design of advanced interfaces and suggest the development of hybrid circuits that couple the strength of nanoelectronics to biological computing components. One of the more attractive materials employed to develop nano-bio hybrid systems is represented by carbon nanotubes (CNTs). CNTs have been at the forefront of nanotechnology and can be manufactured with a very broad range of thermal, electronic, and structural properties and their surface modalities can be adapted very conveniently (Tasis et al. 2006). Due to their outstanding features, CNTs have been quickly recognized as a technology platform for biomedical applications (Sucapane et al. 2008).

In the context of extracellular electrodes, the exploration of nanotechnologies, and specifically of CNT, captured a significant momentum (Keefer et al. 2008). In fact, although inadequate for accurate recording and stimulation access at the subcellular level, extracellular electrodes fulfill other equally important prerequisite for in vivo applications, including the penetration of mechanically tough connective tissue (e.g., the dura mater), access to neurons whose size is much smaller than that of a glass pipette tip, need of developing chronic electrical stimulation and recording without electrochemical cellular insults, as well as the growing demand in designing electrodes with increased sensitivity without apparent loss in selectivity. Mechanical stability is particularly relevant not only in the context of acute in vivo animal experiments, but also in cases, where chronic unrestrained experimental animal preparations are used or during neurosurgical procedures. In addition, recording or stimulating electrical activity from accurately spaced CNS sites in parallel, crucial for topographic functional analysis, avoid considering any alternative technique.

In this chapter, we review the recent advances in employing CNT as electrically conductive substrates for neuronal growth. Emphasis is given to the use of CNTs as electrical interfaces to individual neurons and, in particular, to the modeling of the CNT–neuron junction properties.

## 2 Carbon Nanotube Electronics

CNTs are cylindrically shaped nanostructures constituted by sheets of graphene rolled up to form hollow tubes. Historically, multi-walled carbon nanotubes (MWNTs) were characterized in 1991, (Iijima 1991) followed, later in 1993, by their single-walled analogous (Iijima and Ichihashi 1993; Bethune et al. 1993). However, evidences for nano-sized carbon tubes trace back to the 1952, when two Russian scientists, Radushkevich and Lukyanovich, published TEM images of nano-sized hollow carbon filaments. Since the original article was written in Russian, that can explain why this work was unrecognized by most of the scientists in the rest of the world. For details on this issue, please consult a recent editorial report about the full history of the nanotubes' discovery (Monthioux and Kuznetsov 2006).

Single-walled carbon nanotubes (SWNTs) possess the simplest geometry, i.e., a rolled-up graphene sheet that is closed by fullerene-like caps. Their diameter is in the order of 0.7–2 nm while their length can reach up to several millimeters. Depending on the orientation of the tube axis with respect to the hexagonal lattice, the structure of a nanotube can be completely specified through its chiral vector, which is denoted by the chiral indices $(n, m)$ (Fig. 1). When the chiral indices $n = m$ nanotubes are metallic and are considered quasi metallic (with a tiny band gap) when $n - m$ is divisible by 3. All other tubes are semiconducting with band gaps of the order of 0.5 eV. SWNTs can be either metallic or semiconducting depending on their geometry; CNTs are to date the only material known to have this unique property (Saito et al. 1992). The ratio of metallic–semiconducting CNTs in a raw sample is about 1/3 of metallic and 2/3 of semiconducting. On the contrary, MWNTs present metallic properties because they are constituted of multiple layers of graphene rolled on themselves.

The large aspect ratio, the electrical properties, and the chemical and thermal stability of individual nanotubes make CNTs promising materials for molecular electronics (Léonard and Talin 2006; Avouris and Chen 2006; Avouris et al. 2007). In SWNT, electrons propagate along the tube axis. In particular, electrical transport in metallic SWNT was found to be ballistic over micron lengths at low biases (Bachtold et al. 2000) while in semiconducting tubes the mean free path is about 700 nm (Fuhrer et al. 2001). Individual metallic SWNTs are able to carry currents with a density exceeding 109 $A\,cm^{-2}$ (Yao et al. 2000). Because of this high current density and also because of their mechanical properties, CNTs (i.e., metallic SWNT and MWNT) have been suggested as interconnect materials in silicon nanoelectronics (Graham et al. 2005).

On the other hand, semiconducting SWNTs represent promising materials in electronics to replace (or in combination with) silicon-based devices. While the

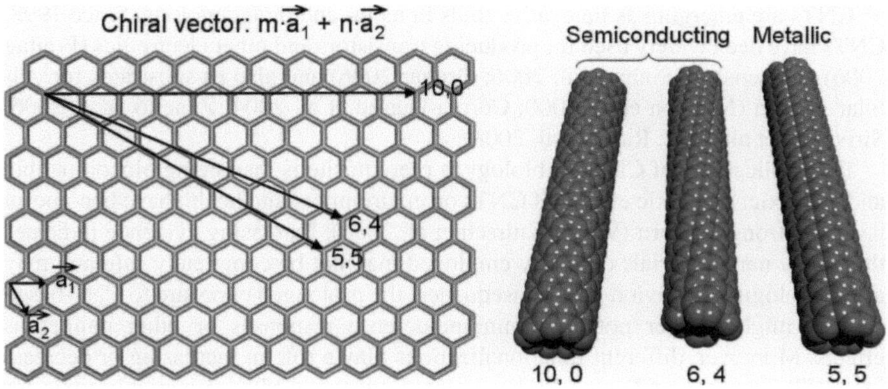


**Fig. 1** Description of the chiral vectors of different nanotubes: zigzag (10,0), chiral (6,4), and armchair (5,5) nanotubes

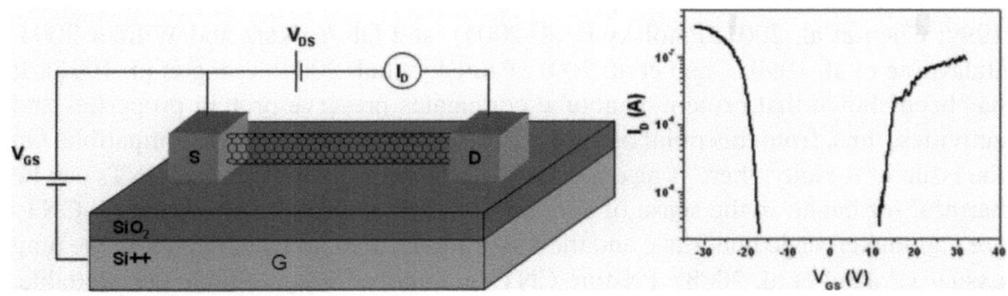


**Fig. 2** Schematic representations of a CNT field-effect transistor (CNT-FET) and its ambipolar behavior. *S* source, *D* drain, *G* gate, $V_{GS}$ gate voltage, $I_D$ device current, $V_{GS}$ drain voltage

resistance in a nanotube with metallic electrical properties does not change by applying another potential (*gate* voltage), the current in a semiconducting nanotube can be tuned by applying external stimulations. This behavior led to the fabrication of CNT field-effect transistors (CNT-FETs; Martel et al. 2001; Tans et al. 1998). The basic structure of CNT-FET involves two metal electrodes designated as *source* and *drain* connected by a semiconducting nanotube. A third electrode, the *gate* is separated from the nanotube by a thin insulator film (Fig. 2 left). At the initial stage, if no potential is applied between the gate and the source, no current crosses the device and this describes the *OFF* state. In a p-type CNT-FET, if a negative gate voltage is applied at a certain threshold, hole current appears in the nanotubes. Reciprocally, for an n-type transistor, an electron current flows when a positive gate voltage is applied and the voltage exceeds the threshold (Fig. 2 right). Similarly to complementary metal–oxide–semiconductor (CMOS) technology, the ON/OFF switching is obtained by sweeping the gate voltage.

CNTs are emerging as innovative tools in nano- and biotechnology. Since 1998, CNTs have been widely used for producing transistors and other electronics (Freitag 2006) and sensors (Sinha et al. 2006; Gruner 2006) and also as substrates for cellular growth (Mattson et al. 2000; Correa-Duarte et al. 2004; Zanello et al. 2006; Sirivisoot et al. 2007; Razal et al. 2008).

For applications of CNTs in biology, a prerequisite is they being biocompatible and nontoxic. The toxic effects of CNTs on environment and health have become an issue of strong concern (Wörle-Knirsch et al. 2006). Increasing evidence indicates that many nanomaterials currently employed may not be completely safe and may affect biological behavior. As a consequence, the prolonged exposure to CNT-based devices might trigger potential immune-defense responses or other dangerous effects. Moreover, different functionalizations play a role in increasing or decreasing CNTs' toxicity and some studies pointed out that CNTs are variably toxic for cells (Sato et al. 2005; Kagan et al. 2006; Sayes et al. 2006; Magrez et al. 2006) while others reported the absence of a particular toxicity (Dumortier et al. 2006). The interfacing of CNTs with biological systems has been also focused on the interactions between CNTs and relevant molecules, such as peptides (Dieckmann et al. 2003; Katz and Willner 2004), proteins (Katz and Willner 2004; Balavoine et al. 1999; Chen et al. 2001; Patolsky et al. 2004), and DNA (Katz and Willner 2004; Balavoine et al. 1999; Chen et al. 2001; Patolsky et al. 2004; Tsang et al. 1997). It has been shown that protein–nanotube conjugates preserve protein properties and activities; thus, from this point of view, CNTs can be regarded as biocompatible. On the issue of toxicity, there is also a general agreement on the fact that CNTs can be harmful for health in the sense of similarly to other micro- or nanoparticles, CNTs are nanometer-scale materials, and their size allows them to enter deeply in the lung tissues (Poland et al. 2008). Pristine CNTs are neither water soluble nor wettable, and so they are extremely difficult to eliminate from the body. Recently, Lam et al. demonstrated that CNTs were responsible for inflammation, epithelioid granulomas (microscopic nodules), fibrosis, and biochemical and toxicological changes in the lungs (Lam et al. 2004). This study also revealed that SWNTs were more toxic than quartz particles. It should be noted, however, that this material appears to be biocompatible, since systemic administration of CNTs does not cause any detectable toxic effects (Liu et al. 2008).

In 2000, a study realized by Mattson and coworkers suggested that CNTs could be used as substrates for neuronal growth. After this first report, several groups developed single neurons or cultured neuronal circuits using as substrate a CNT thin film (Hu et al. 2004; Hu et al. 2005; Lovat et al. 2005; Mazzatenta et al. 2007; Gheith et al. 2006; Pappas et al. 2007; Wang et al. 2006; Matsumoto et al. 2007; Gabay et al. 2007). The electrical properties of CNTs as well as their biocompatibility (at least when immobilized on surface) make these materials the perfect candidates for neuronal growth and also for development of implantable devices displaying peculiar properties in interfacing with neuronal electrical activity, and thus CNTs could serve as an extracellular scaffold, sustaining and interacting with neurite growth and signaling, with a wide range of applications in neurology health care.

## 3 Carbon Nanotubes as Electrically Conductive Cell Growth Substrates

In the following, we review some electron transport properties of CNTs. Although simplified, these considerations are probably sufficient to address the phenomena at the interface between a (mesoscopic) dense random dispersion of CNTs and the membrane of a neuron, as in Fig. 3 (see Mattson et al. 2000; Lovat et al. 2005; Liopo et al. 2006; Mazzatenta et al. 2007).

In the case of a mesoscopic mixture of CNTs, randomly deposited or patterned on a planar substrate employed as a cell-growth surface, an intricate *meshwork* of

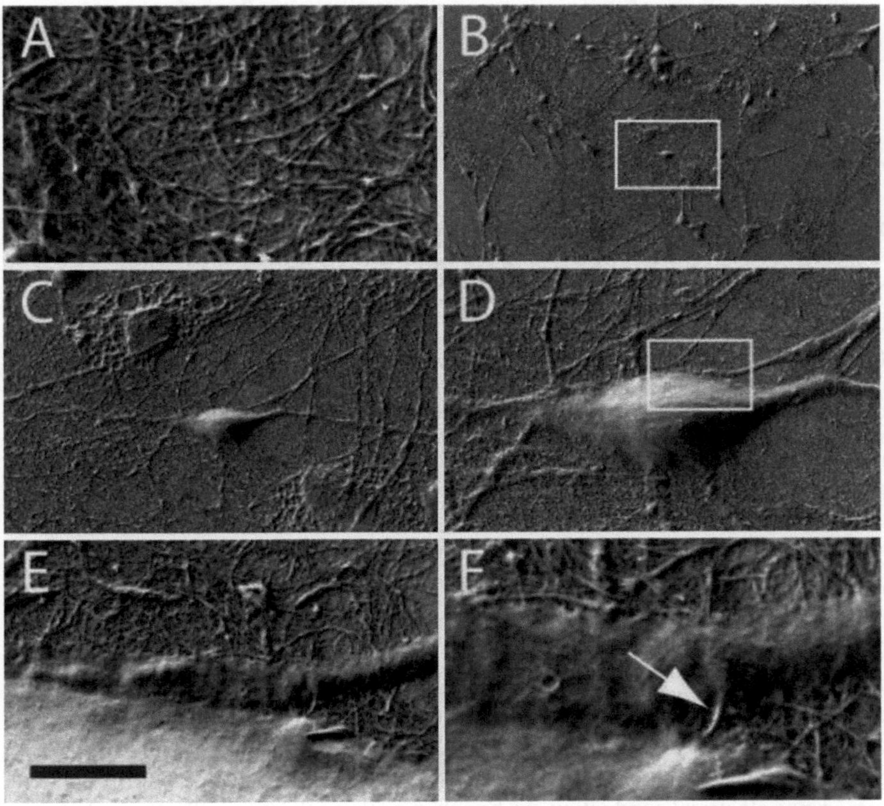


**Fig. 3** Scanning electron microscopy images of cultured neurons growing on CNTs. At very high magnification (**a**), the intricate structure of the nanotube meshwork obtained from the CNT dispersion is apparent. Neurons grow and develop ex vivo on such a substrate for several weeks, reorganizing into functional networks (**b–d**). At the subcellular level (**e–f**), tight contacts between cell membranes and individual nanotube bundles are clearly identified, suggesting a very intimate biophysical coupling. Scale bar: (**a**) 1 μm, (**b**) 200 μm, (**c**) 25 μm, (**d**) 10 μm, (**e**) 2 μm, (**f**) 450 nm. © 2007 Society for Neuroscience, reproduced with permission from Mazzatenta et al. (2007)

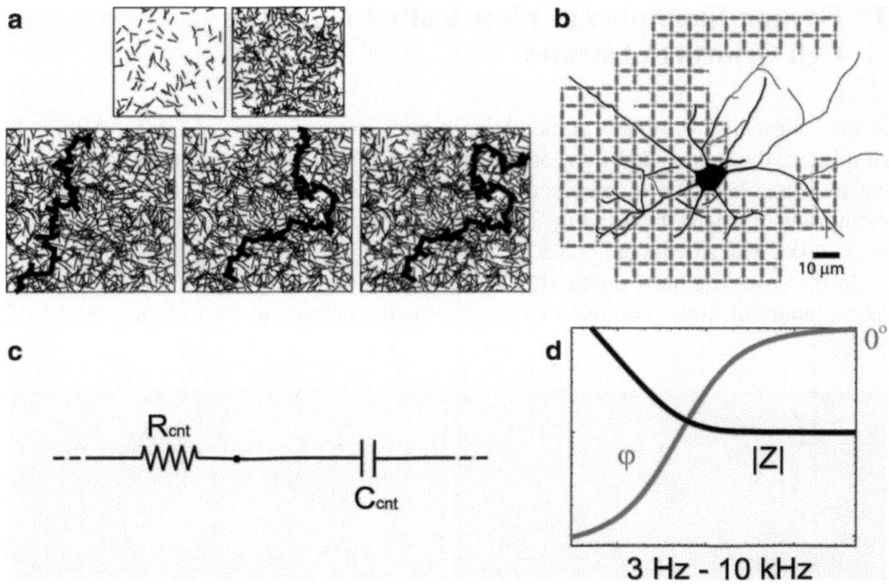


**Fig. 4** Model of CNTs as a cell-growth substrate. Random dispersions of computer-generated small segments on a finite area (**a**) mimic CNT surface coating for cell growth. For increasing dispersion density, a dense meshwork of elements appears, and the network percolation limit for electrical conductivity is readily overcome. Thus, between two points of the (simulated) substrate, there exist a *chain* of conductive elements, forming a resistive electrical path (outlined in *dark*). Individual neurons cultured over CNT layers are thought coupling to a heterogeneous 2-dimensional resistive network, sketched in (**b**) as a resistive grid. Electrochemical properties of the interface between the CNT substrate and the electrolyte are instead mostly capacitive and described by a standard electrochemical impedance model (**c**); $R_{cnt}$, CNT equivalent resistance; $C_{cnt}$, CNT equivalent capacitance. This description is equivalent to a linear impedance Z, whose magnitude $|Z|$ and phase $\varphi$ have been sketched qualitatively in the frequency domain (**d**)

nanotubes with semiconducting and metallic properties is usually obtained, as visualized by electron microscopy in Fig. 3a (Radosavljevic et al. 2001; Georgakilas et al. 2002; Snow et al. 2003). Because of the high surface density, a large amount of tubes with metallic properties are randomly dispersed and oriented among the others. Therefore, a *percolating* network of electrical pathways between any two points of the mesh is likely to emerge (Snow et al. 2003). Then, even with relatively small probability of contacts between two neighboring tubes with metallic properties, long-range electrical conduction can be achieved through intermediate steps between any two points of the mesh (Kirckpatrick 1972). This has been graphically illustrated in Fig. 4a, where a simple geometrical computer model of randomly dispersed segments was simulated and analyzed. As in a *random graph* (Bollobás 2001), two neighboring *nodes* are connected if the corresponding "black needles" intersect each other. Then, for sufficiently large network sizes, a path between any two points almost certainly exists in the simulated nanotube dispersion (Fig. 4a).

The sketch of Fig. 4b summarizes the conceptual simplification that we adopt here for describing the effective electron transport properties of the CNT meshwork, deposited on a planar substrate, constituting a growth surface for ex vivo networks of dissociated neurons. The branched outline depicted in Fig. 4b is a 2-dimensional reconstruction of the actual morphology of a cultured hippocampal neuron, dissociated from rat brain and developing in vitro on an adhesion substrate made of nanotubes as in Fig. 3 (see also Lovat et al. 2005). We hypothesize that most of the neuronal cellular processes grow in close proximity of the heterogeneous meshwork, largely composed of resistive links. While such a cartoon accounts for the lateral conduction within the meshwork, it leaves unspecified the details related both to the substrate-electrolyte and to the substrate-neuronal membrane physical interfaces (modeled in Fig. 4c–d). Indeed, charge carriers in an electrolyte are not free electrons as in the nanotubes.

## 3.1 *Carbon Nanotube Electrodics*

The electrical properties of the interface CNT–electrolyte have been characterized, mostly in the context of biosensing and electrodics (Czerw et al. 2006; Crespo et al. 2008), by electrochemical impedance spectroscopy (see Girault 2004, for an introduction). Although most of the current literature focuses on the CNTs' excellent amperometric properties when functionalized with polymers and enzymes (see, e.g., Sinha et al. 2006; Zhang et al. 2007; Male et al. 2007; Muguruma et al. 2008; Roy et al. 2008; Crespo et al. 2008), very few investigators fully realized the great potential of CNTs as novel materials for (neuroprosthetics) electrodes (Phely-Bobin et al. 2006; Gabay et al. 2007; Mazzatenta et al. 2007; Li and Andrews 2007; Keefer et al. 2008). The nanoscale sizes of individual CNTs as well as their considerable mechanical rigidity suggest replacing large metal electrodes with nanostructures comparable in size with subcellular neuronal details (*see,* e.g., Fig. 3) while exploiting the electrical properties of the large exposed surface (i.e., electrical capacitance).

Since CNTs are inert in saline solutions, electrical flow at the interface between a saline solution and CNTs ideally occurs as a capacitive current only, as charge carriers in the solutions (i.e., ions) and charge carriers in the CNT (i.e., electrons) cannot be directly exchanged. Instead, at the equilibrium, a redistribution of charge carriers in both the CNT and the electrolyte occurs at the interface as in a capacitor, forming what is called the electric double layer (see Geddes 1972; Robinson 1968; Bockris et al. 2000). The electric double layer approximates literally an electrolytic capacitor, and for CNT dispersions with high surface area it results into a large effective capacitance, compared, e.g., to gold electrodes. Faradaic reactions at the interface are limited to oxygen chemisorptions (Girault 2004), causing conduction leakage current passing through the interface at very low frequencies and large electric fields. Figure 4c–d sketches an equivalent electrical circuit model, describing effectively the electrochemical impedance at the CNT–electrolyte interface. Such a model is employed in the next sections, when a biophysical model of the CNT–neuron junction is introduced.

Because the surface charge redistribution (both in the electrolyte and in the CNT meshwork) occurs in conducting media, a characterization of the physical system also in terms of distributed parameters could be more appropriate. However, on the basis of the electrochemical considerations employed in metal–electrolyte interfaces (Robinson 1968), we decided to focus only on the frequency range that characterizes (extracellular) neuronal signaling (i.e., 100 Hz–10 kHZ). Within such a bandwidth, the model of Fig. 4c, d is adequate to describe the signal transduction at the interface, and we further consider the frequency-independent resistance and capacitance as correct, lumped descriptions of our system.

## 3.2 The CNT–Neuron Junction

Based on the work of Grattarola and Martinoia (1993), we now introduce a method useful to describe the biophysical and electrochemical properties of the interface between CNTs and neurons. As already mentioned for the CNT–electrolyte interface (Fig. 4c, d), such an approach consists in the formulation of linear electrical circuit models composed of capacitors and resistors, which are equivalent to the differential equations describing the temporal evolution of the extracellular electrical variables of interest. This description shares obvious similarities with the classic approach employed for the electrical excitability in neurons (Hodgkin and Huxley 1952; Fig. 7b). The exploration and applications of these concepts to the description of the interface between metal and the neuronal membrane gained renewed interest in the last few years, thanks to the increasing availability of multisite in vitro electrophysiological chronic recordings from cultured networks of dissociated neurons in neuroelectronic studies (Fig. 5a–b) (Rutten 2002 and references therein). In fact, the specific features of the voltage waveforms, detected extracellularly by means of substrate metallic microelectrodes (Fig. 5d), could be related to the intracellular action potentials (Fig. 5c) "filtered" by the biophysical and electrochemical properties of the cell–electrode coupling and of the electrolyte (Grattarola and Martinoia 1993; Martinoia et al. 2004).

Figure 5 reports the intracellular and extracellular recordings of the spiking activity, which emerges spontaneously in primary cultures of neurons (Marom and Shahaf 2002). Simultaneous patch recordings (Fig. 5a) reveal the typical intracellular signals expected during action potentials, and unveil subthreshold, network-driven synaptic activity that underlies the spontaneous occurrence of bursts of action potentials (Fig. 5c). On the other hand, the same electrical activity, detected extracellularly by thin-film substrate microelectrodes (Fig. 5b), allows enhanced spatial resolution at the expenses of the temporal resolution: indeed, only suprathreshold (i.e., spiking) activity is detected extracellularly (Fig. 5b, d). On the basis of our current knowledge, CNT-based substrate electrodes are expected to behave qualitatively as metallic extracellular electrodes, with unique and unparalleled physical properties. Through our biophysical modeling approach (Fig. 8), one can quantitatively appreciate how these features boost signal-to-noise ratio (Fig. 8c) and record-

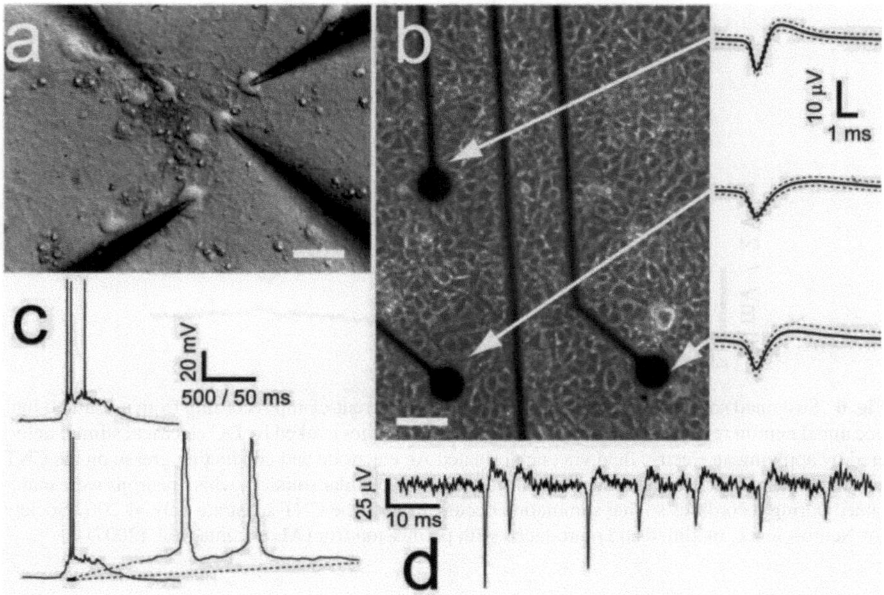


**Fig. 5** Chronic detection of spontaneous electrical activity in neuronal cultures. Primary neuronal cultures (**a–b**) can be maintained under healthy conditions for several weeks, growing on arrays of substrate planar extracellular metallic electrodes (**b**). Extracellularly detected spikes (**b, d**) display the stereotypical features of extracellular recordings. As apparent from the sample traces (**b, d**), extracellular signals considerably differ from those obtained by means of simultaneous intracellular multipatch recordings (**a, c**). Horizontal calibration: 25 μm in **a**, and 50 μm in **b**

ing selectivity. In addition, there are key evidences indicating that CNT-based materials display a peculiar signal coupling resembling an intracellular (i.e., patch) and not extracellular access to the intracellular membrane potential (Liopo et al. 2006; Mazzatenta et al. 2007; see also Schoen and Fromherz 2007). Such a coupling cannot occur at the interface between macroscopic metal electrodes and neuronal membranes due to the generally smooth surface and lack of nanostructures. This is shown in Fig. 6, where a train of sustained action potential is evoked by CNT-mediated electrical stimulation, reminiscent of a sustained direct intracellular current flow. Although interpretation of these data requires careful discussion and an assessment on the details of electrophysiological technique (see Mazzatenta et al. 2007 for a discussion), such evidences point out that intimate mechanical proximity between bundles of CNTs and the neuronal membrane (Fig. 3f) might correlate to an intracellular-like access of the cytosolic cell compartments.

For the general character of our considerations, 2-dimensional morphologies of a cultured neuron (Fig. 7a), as reconstructed and digitized from microscope images through a basic camera-lucida tracer (freely available at the Matlab Central Web site, file id: 8336, The Mathworks, Natick, MA), can be reduced to a 1-dimension

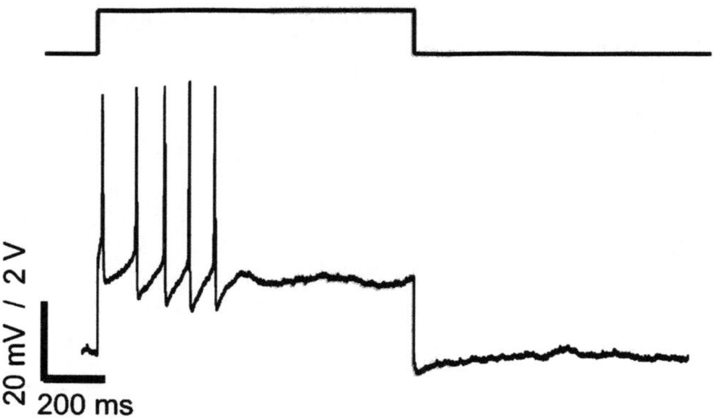


**Fig. 6** Sustained spiking evoked by CNT stimulation. Current-clamp recording from a cultured hippocampal neuron reveals sustained action potential discharges evoked by DC electrical stimuli delivered by applying an electric field via unchlorinated Ag electrode and conducting grease on the CNT substrate mesh. The electrode was isolated from the extracellular solution, where neurons were maintained during recordings so that stimulation occurs through the CNT substrate only. © 2007 Society for Neuroscience, modified and reproduced with permission from Mazzatenta et al. (2007)

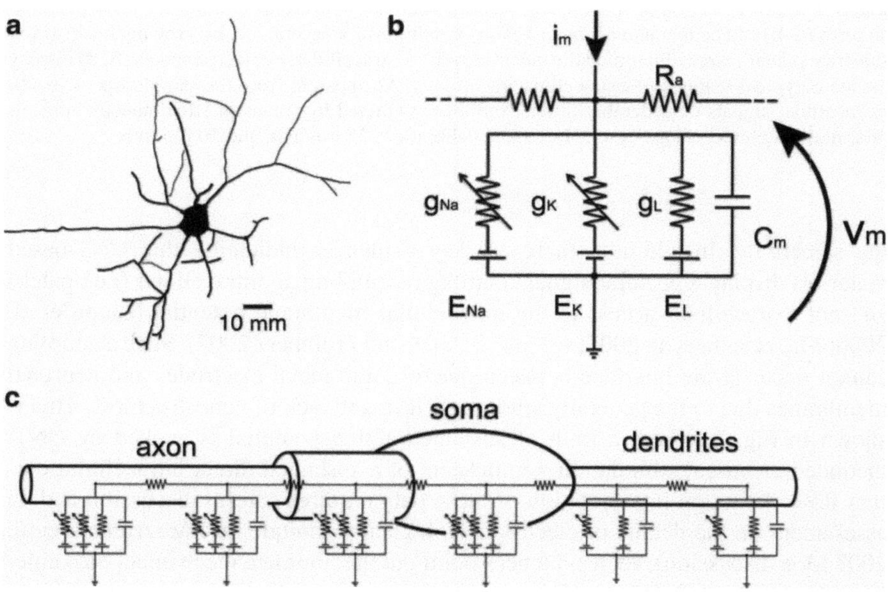


**Fig. 7** Equivalent circuit of a multicompartmental conductance-based model neuron. Complex 2D morphologies of cultured hippocampal neurons (**a**) can be reduced into 1D nonlinear cable structures and compartmentalized by the standard Hodgkin–Huxley approach (**b–c**). $C_m$ membrane capacitance, $E_{Na}$ $Na^+$ reversal potential, $E_K$ $K^+$ reversal potential, $E_L$ leak reversal potential, $g_{Na}$ $Na^+$ conductance, $g_K$ $K^+$ conductance, $g_L$ leak conductance, $R_a$ axial cytoplasmic resistance, $V_m$, membrane potential

cable (Fig. 7c). For the sake of simplicity, such morphology is further compartmentalized into a series of heterogeneous segments that are electrically coupled to each other through the axial ion flow, occurring within the neuronal cytoplasm. Both passive and excitable electrical properties of each compartment are assumed to arise from voltage-gated ion conductances (Dayan and Abbott 2001; Traub and Miles 1991) and modeled according to standard electrical equivalents, as originally proposed by Hodgkin and Huxley (1952) (Fig. 7b, c).

In the model of Fig. 7c, the membrane potential $V_{m,j}$ and the ion current at the *j*th compartment can be expressed in terms of ion currents:

$$i_{m,j} = C_m \cdot \frac{dV_{m,j}}{dt} + i_{\mathrm{Na},j} + i_{K,j} + \ldots + i_{\mathrm{leak},j} + \frac{2 \cdot V_{m,j} - V_{m,j-1} - V_{m,j+1}}{R_a}. \quad (1)$$

Equation (1) follows from cable theory (Dayan and Abbott 2001) and it is derived imposing the charge conservation across a patch of neuronal membrane, characterized by capacitance $C_m$ and axial cytoplasmic resistance $R_a$. We further indicated by $i_{\mathrm{Na}}$, $i_{\mathrm{K}}$, and $i_{\mathrm{leak}}$ the transmembrane ion currents. Each of these currents is selective to different ion species (e.g., $Na^+$, $K^+$, $Cl^-$) and it is characterized by maximal conductances ($g_{\mathrm{Na}}$, $g_{\mathrm{K}}$, and $g_{\mathrm{leak}}$) and apparent reversal potentials ($E_{\mathrm{Na}}$, $E_{\mathrm{K}}$, and $E_{\mathrm{leak}}$). Reversal potentials are modeled as ideal voltage sources and their values satisfy the Nernst–equilibrium relationship for the corresponding ion species in solution (Fig. 7b) (Dayan and Abbott 2001). As in the Hodgkin–Huxley model (Hodgkin and Huxley 1952), first-order kinetic schemes account for the instantaneous fractions of voltage-gated sodium and potassium channels:

$$\begin{aligned} i_{\mathrm{Na},j} &= g_{\mathrm{Na}} \cdot m_j^{\,3} \cdot h_j \cdot \left(V_{m,j} - E_{\mathrm{Na}}\right) \\ i_{\mathrm{K},j} &= g_{\mathrm{K}} \cdot n_j^{\,4} \cdot \left(V_{m,j} - E_{\mathrm{K}}\right) \\ i_{\mathrm{leak},j} &= g_{\mathrm{leak}} \cdot \left(V_{m,j} - E_{\mathrm{leak}}\right) \end{aligned} \quad (2)$$

This model can be implemented and computer simulated by standard neuronal simulators, such as the NEURON environment (Carnevale and Hines 2006), and reproduces some of the electrophysiological features of real-cultured neurons, such as the resting membrane potential at −70 mV, a membrane time constant in the range 10–20 ms, an input resistance of 100–130 MΩ, a rheobase current of 30–50 pA, a spike overshot of 20–30 mV, as well as the dendritic backpropagation of action potentials (Schaefer et al. 2003).

Similarly to the ion flows occurring across the neuronal membrane, the electric fields and current flows at the CNT–neuron interface can be described in terms of equivalent circuit models (Fig. 8a–b). The electrochemical model first discussed in Fig. 4c, accounting for the CNT–electrolyte interface, was incorporated and the values of the resistance $R_{\mathrm{cnt}}$ and capacitance $C_{\mathrm{cnt}}$ were varied to account for either metal or CNT properties (Fig. 8c). The (intimate) coupling between the substrate and the neuronal membrane is modeled in terms of a "seal" resistance and of a

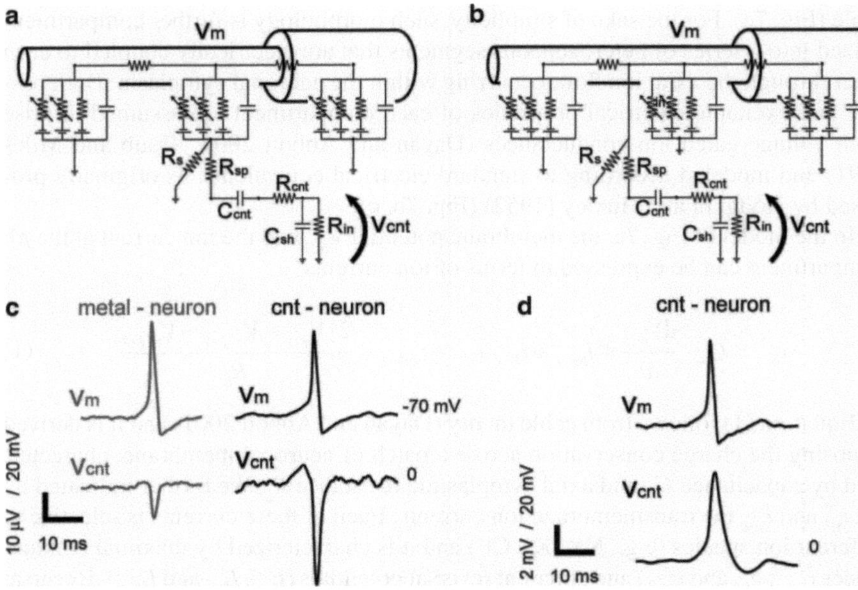


**Fig. 8** Equivalent circuit models for the CNT neuron. A biophysical mathematical model combining voltage-dependent cell excitability with the interfacing of the neuronal membrane to CNT-based electrodes, as implemented in NEURON. Two versions of the same model were defined. The first (**a**) assumes that an extracellular coupling always occurs between the neuronal membrane and the CNTs, as in metal microelectrodes. As expected, simulations of this model predict a much larger signal-to-noise ratio, compared to a metal microelectrode (**c**). The second model hypothesized that bundles of CNTs penetrate the lipid double layer and become exposed to the cytosolic environment (**b**). Under these conditions, the model predicts a response amplitude two orders of magnitude larger and anticipates a qualitatively different signal shape, intriguingly similar to the intracellular membrane potential (**d**). $C_{cnt}$, CNT equivalent capacitance; $C_{sh}$, amplifier shunt capacitance; $R_{in}$, amplifier input resistance; $R_s$, seal resistance; $R_{sp}$, spread resistance; $R_{cnt}$, CNT equivalent resistance; $R_s$, seal resistance; $V_{cnt}$, CNT potential; $V_m$, membrane potential

"spread" resistance. These components describe the ease of ion currents to couple the extracellular space to the CNT and quantify the electrical shunt to the bulk electrolyte. In Fig. 8a–b, $R_s$ is the sealing resistance between the neuronal membrane and the surface of the CNT layer, and it accounts for how much the neuronal membrane is in proximity with the CNT electrode. Depending on the size of the electrolyte cleft between the neuronal membrane and the nanotubes, this resistance assumes distinct values, increasing with decreasing distance. $R_s$ depends, therefore, only on the resistivity $\rho_s$ of the electrolyte (i.e., $\rho_s = 70\ \Omega\text{cm}$) and on geometrical parameters (Fromherz 2002); thus, $R_{seal} = \rho_s \cdot \delta / d$, $\delta$ is surface overlapping coefficient and $d$ is thickness of the electrolyte cleft. Because at a mesoscopic scale the surface of the CNT substrate appears as a fractal landscape (Fig. 3a), the thickness $d$ of the cleft of electrolyte and the surface overlapping coefficient $\delta$ (Grattarola and Martinoia 1993) that quantifies the contact area between the neuron and the CNT should be interpreted only as average parameters. Similarly, $R_{sp}$ depends on the electrolyte resistivity and on geometrical parameters. As opposed to conventional extracellular coupling of the membrane of cultured neurons with microtransducers integrated in the culture substrate (Grattarola and Martinoia 1993; Martinoia et al. 2004), $R_s$ is probably strongly affected by the extremely intimate contact reported by our previous studies (Mazzatenta et al. 2007). As anticipated in the model of Fig. 8b, d, pinching and complete penetration of CNTs in the membrane lipid layers are certainly possible, although so far only based on the interpretation of TEM microscopy images (Cellot et al. 2008). In such a case, the biophysical model predicts a signal transduction fundamentally different from metal electrodes, reminiscent of an intracellular, though noninvasive, recording.

The full model implementation is available online at the ModelDB database for the NEURON simulation environment (https://senselab.med.yale.edu/modeldb/ShowModel.asp?model=112086). The great advantage of employing a standard neural simulator as NEURON, conceived specifically for biophysically detailed models of neuronal excitability, is represented by the more natural implementation of individual ion current kinetics accounting for neuronal excitability and signal propagation even in complex morphologies (Traub and Miles 1991). There is in fact no need to invoke time-invariant electrical equivalent circuits (Martinoia et al. 2004; Storace et al. 1997; Chua 1980) for the convenience of specification under standard electrical circuit simulators (e.g., HSPICE, NGSPICE, etc.). Arbitrarily complex morphologies and network architectures (Migliore et al. 2006; Markram 2006) can be, then, immediately implemented. A large set of available channel kinetics and neuronal point process can be also included in the model to account for a more realistic biophysical description (e.g., from the SenseLab ModelDB – http://senselab.med.yale.edu/modeldb; Hines et al. 2004). Furthermore, the availability of optimized numerical routines represents an explicit advantage for accuracy and computation speed. Finally, the recent versions of the NEURON simulation environment allows one to simulate additional linear (and nonlinear) circuit models by no effort (Gold et al. 2006; Carnevale and Hines 2006), making the modeling of extracellular signal recording and stimulation very natural in terms of electrical equivalent models of the neuron–nanotube interface.

### 3.3 Toward a Deeper Understanding of CNT–Neuron Junctions

At the present time, our understanding of the biophysics of the CNT–neuron junction is preliminary and the description of the electrical properties of nanotubes' meshwork outlined in the previous section is a simplified one. The attempts, from our group and from others (Gheith et al. 2006; Liopo et al. 2006; Mazzatenta et al. 2007), to engineer CNTs as cell-culturing substrates are also elementary (see also Gabay et al. 2007). In addition, the actual electron transport in a dense CNT meshwork is certainly depending on additional details (Snow et al. 2003). In particular, electrostatic charges and surface adsorption of various molecules, and their impact on the one-dimensional electron conduction in individual CNTs, have been so far neglected but are expected to play important roles. Indeed, environmental electrostatic influences closely resemble the voltage-dependent gating of semiconductor conductance (i.e., the field effect), exploited in metal–oxide–semiconductor (MOS) transistors (i.e., MOS-FETs). This has been almost exclusively studied under the perspectives of engineering new generations of CNT-based transistors (Avouris et al. 2005). There, the semiconducting properties of SWNTs and their modulation by external electric fields are crucially placed and represent the basis of the transistor effect.

Given the nanoscale of the CNT meshwork details, do time-varying ion currents influence CNT lateral conduction, when occurring at the neuron–CNT interface? A more in-depth description of semiconductive phenomena and CNT–CNT resistive junctions will soon become imperative. In particular, local extracellular electric fields, induced by the neuronal membrane during an action potential (Rall 1962), might directly affect the instantaneous lateral conduction in the nanotube substrate (Kim and Kim 2007). If this is the case, then one could envision the activation of spike-triggered percolation networks on a mesoscopic scale, with electrical shunting of proximal neurons that fired synchronously. Only additional experiments and the advance in our understanding of the CNT/neuron biophysics in culture clarify whether these scenarios are relevant and beneficial for future neuroprosthetics.

Another central issue that has been explored in the context of biosensing applications, but never related so far to the neurophysiology of neurons in contacts with CNTs, is represented by the nonspecific binding of proteins to the nanotube surface (Chen et al. 2003). This phenomenon can modulate the electrical transport properties of isolated CNTs, similarly to the gating electric field discussed above. We certainly expect that future developments and investigations will ultimately focus on the understanding of nanotubes in close proximity with a living biological tissue matrix. Cells are known to synthesize their own extracellular adhesion proteins (Lutolf and Hubbell 2005), which may interact directly with the CNT substrate and affect its electrical properties. In addition, excitable cells like neurons might even (endogenously) induce selective protein adsorption or release from the substrate or from previously engineered substrates (e.g., loaded with neuroactive amino acid compounds) (Mattson et al. 2000). These phenomena might certainly have a substantial impact on the substrate lateral conductivity. The manipulation of this class of phenomena is exploited for nano-neuroengineering applications (Patolsky et al. 2006), opening exciting perspectives for future applied clinical research applications (Silva 2006; Benabid et al. 2005; Llinás et al. 2005).

## 4 Conclusions

Characterizing the electrical properties of the interface between CNTs and neurons help in elucidating the potential of CNTs as novel materials for (neuroprosthetics) electrodes. More specifically, innovative aspects in CNT/neuron interactions, such as the presence of signal transduction reminiscent of an intracellular, though noninvasive recording, should be specifically addressed to develop new (nano) technological tools to probe the brain.

**Acknowledgments** Authors acknowledge financial support from the Italian Ministry of University and Research (Cofin and FIRB), from the École Polytechnique Fédérale de Lausanne EPFL, from the European Commission (NEURONANO-NMP4-CT-2006-031847), and from the "Stoicescu" grant. M.G. and L.G. are grateful to Prof. H. Markram, C. Petersen, and S. Martinoia for helpful discussions and to S. Garcia and K. Antoniello for assistance.

## References

Avouris P, Radosavljevic M, Wind SJ (2005) Carbon Nanotube Electronics and Optoelectronics, in "Applied Physics of Carbon Nanotubes, Fundamentals of Theory, Optics and Transport Devices", (Rotkin SV, Subramoney S, eds.), Springer Berlin, 227–49.

Avouris P, Chen J (2006) Nanotube electronics and optoelectronics. Materials Today 9, 46–54.

Avouris P, Chen Z, Perebeinos V (2007) Carbon-based electronics. Nature Nanotech. 2, 605–615.

Bachtold A, Fuhrer MS, Plyasunov S, Forero M, Anderson EH, Zettl A, McEuen PL. (2000) Scanned probe microscopy of electronic transport in carbon nanotubes. Phys Rev Lett. 84(26 Pt 1): 6082–5.

Balavoine F, Schultz P, Richard C, Mallouh V, Ebbesen TW, Mioskowski C (1999) Helical Crystallization of Proteins on Carbon Nanotubes: A First Step towards the Development of New Biosensors. Angew. Chem. Int. Ed. 38:1912–1915.

Bollobás B (2001) Random graphs (2nd ed.), Cambridge University Press.

Benabid AL, Wallace B, Mitrofanis J, Xia C, Piallat B, Fraix V, Batir A, Krack P, Pollak P, Berger, F (2005) Therapeutic electrical stimulation of the central nervous system. C R Biol 328(2): 177–86.

Bethune DS, Kiang CHM, de Vries S, Gorman G, Savoy R, Vazquez, J, Beyers R (1993) Cobalt-catalyzed Growth of Carbon Nanotubes with Single-atomic-layer Walls. Nature 363:605–7.

Bockris J, Reddy AKN, Gamoa-Aldeco M (2000) Modern Electrochemistry 2A: Fundamentals of Electrodics, 2nd ed., Kluwer Academic, New York (USA).

Carnevale NT, Hines ML (2006) The NEURON book. Cambridge University Press, Cambridge, UK.

Chen RJ, Bangsaruntip S, Drouvalakisdagger KA, Wong Shi Kam N, Shim M, Li Y, Kim W, Utzdagger PJ, Dai H (2003) Noncovalent functionalization of carbon nanotubes for highly specific electronic biosensors. Proc. Natl. Acad. Sci. U S A 100(9):4984–9.

Chua LO (1980) Device modeling via basic nonlinear circuit elements. IEEE Trans. Circuits Syst. 27:1014–44.

Chen, RJ, Zhang, Y, Wang, D, Dai, HJ. (2001) Noncovalent Sidewall Functionalization of Single-Walled Carbon Nanotubes for Protein Immobilization. Am. Chem. Soc. 123:3838–9.

Cellot, G, Cilia, E, Cipollone, S, Rancic, V, Sucapane, A, Giordani, S, Gambazzi, L, Markram, H, Grandolfo, M, Scaini, D, Gelain, F, Casalis, L, Prato, M, Giugliano, M and Ballerini, L. (2009) Carbon nanotubes might improve neuronal performance by favouring electrical shortcuts. Nat Nanotechnol 4:126–133.

Correa-Duarte M, Wagner N, Rojas-Chapana J, Morsczeck C, Thie M, Giersig M (2004) Fabrication and Biocompatibility of Carbon Nanotube-Based 3D Networks as Scaffolds for Cell Seeding and Growth. Nano Letters 4 (11):2233–6.

Crespo GA, Macho S, Rius FX (2008) Ion-Selective Electrodes Using Carbon Nanotubes as Ion-to-Electron Transducers. Analytical Chem. 80(4):1316–22.

Czerw R, Edell D, Farrell B, Fooksa R, Phely-Bobin T, Robblee L, Tiano T (2006) Carbon Nanotube Based Electrodes for Neuroprosthetic Applications. Material Res. Soc. Symposium Proc. 926:19–24, Warrendale, Pa. USA.

Dayan P, Abbott LF (2001) Theoretical Neuroscience: Computational and Mathematical Modeling of Neural Systems. The MIT Press, Cambridge.

Dieckmann, GR, Dalton, AB, Johnson, PA, Razal, J, Chen, J, Giordano, GM, Munoz, E, Musselman, I H, Baughman, RH, Draper, RK. (2003) Controlled Assembly of Carbon Nanotubes by Designed Amphiphilic Peptide Helices. J. Am. Chem. Soc. 125:1770.

Dumortier H, Lacotte S, Pastorin G, Marega R, Wu W, Bonifazi D, Briand JP, Prato M, Muller S, Bianco A (2006) Functionalized Carbon Nanotubes Are Non-Cytotoxic and Preserve the Functionality of Primary Immune Cells. Nano Lett. 6:152–8.

Freitag M (2006) Carbon nanotube electronics and devices. In "Carbon Nanotubes: Properties and Applicatons", O'Connell MJ ed., CRC Press, pp. 83–117.

Fromherz P (2002) Electrical Interfacing of Nerve Cells and Semiconductor Chips. Chem. Phys. Chem. 3(3):276–84.

Fuhrer MS, Forero M, Zettl A, McEuen PL (2001) Ballistic transport in semiconducting carbon nanotubes. AIP Conf. Proc. 591:401–404.

Gabay T, Ben-David M, Kalifa I, Sorkin R, Abrams ZR, Ben-Jacob E, Hanein Y (2007) Electrochemical biological properties of carbon nanotube based multi-electrode arrays. Nanotech 1:035201, doi:101088/0957-4484/18/3/035201.

Geddes LA (1972) Electrodes the Measurement of Bioelectric Events. New York: Wiley-Interscience.

Georgakilas V, Kordatos K, Prato M, Guldi DM, Holzinger M, Hirsch A (2002) Organic Functionalization of Carbon Nanotubes, J. Am. Chem. Soc. 124(5): 760–1.

Gheith MK, Pappas TC, Liopo AV, Sinani V, Shim BS, Motamedi M, Wicksted JP, Kotov NA (2006) Stimulation of neural cells by lateral currents in conductive layer-by-layer films of single-walled carbon nanotubes. Adv Mater 18:2975–9.

Girault HH (2004) Analytical and Chemical Electrochemistry. Marcel Dekker, Inc., New York, 2004.

Giugliano M, Prato M, Ballerini L (2008) Nanomaterial/neuronal hybrid system for functional recovery of the CNS. Drug Discov. Today: Disease Model, doi:10.1016/j.ddmod.2008.07.004.

Gold C, Henze DA, Koch C, Buzsaki G (2006) On the origin of the extracellular action potential waveform: A modeling study. J Neurophysiol 95:3113–28.

Graham AP, Duesberg GS, Seidel RV, Liebau M, Unger E, Pamler W, Kreupl F, Hoenlein W. (2005) Carbon nanotubes for microelectronics? Small 1(4):382–90.

Grattarola M, Martinoia S (1993) Modeling the neuron-microtransducer junction: from extracellular to patch recording. IEEE Trans Biomed Eng 40(1):35–41.

Gruner G. (2006) Carbon nanotube transistors for biosensing applications. Anal. Bioanal. Chem. 2006 Jan;384(2):322–35.

Hines ML, Morse T, Migliore M, Carnevale NT, Shepherd GM (2004) ModelDB: A Database to Support Computational Neuroscience. J. Comput. Neurosci. 17(1):7–11.

Hodgkin AL, Huxley, AF (1952) A quantitative description of membrane current and its application to conduction and excitation in nerve. J Physiol. 117(4):500–44.

Hu, H, Ni, Y, Montana, V, Haddon, RC, Parpura, V. (2004) Chemically Functionalized Carbon Nanotubes as Substrates for Neuronal Growth. Nano Lett. 4:507–11.

Hu H, Ni Y, Mandal SK, Montana V, Zhao B, Haddon RC, Parpura V (2005) Polyethyleneimine functionalized single-walled carbon nanotubes as a substrate for neuronal growth. J. Phys. Chem. B 109:4285–9.

Iijima S (1991) Helical microtubules of graphitic carbon. Nature 354, 56–58.

Iijima S, Ichihashi T (1993) Single-shell carbon nanotubes of 1-nm diameter. Nature 363:603–5.

Kagan VE, Tyurina YY, Tyurin VA, Konduru NV, Potapovich AI, Osipov AN, Kisin ER, Schwegler-Berry D, Mercer R, Castranova V, A.A. Shvedova (2006) Direct and indirect effects of single walled carbon nanotubes on RAW 264.7 macrophages: Role of iron. Toxicol. Lett. 165:88–100.

Katz E, Willner I (2004) Biomolecule-Functionalized Carbon Nanotubes: Applications in Nanobioelectronics Chem. Phys. Chem. 5:1084.

Keefer KW, Botterman BR, Romero MI, Rossi AF, Gross GW (2008) Carbon nanotube coating improves neuronal recordings. Nature Nanotech. 3:434–9.

Kim WY, Kim KW (2007) Carbon nanotube, graphene, nanowire, and molecule-based electron and spin transport phenomena using the nonequilibrium Green's function method at the level of first principles theory. J. Comp. Chem. 10.1002/jcc.20865.

Kirckpatrick S (1972) Percolation and Conduction. Rev. Mod. Phys. 45(4):574–88.

Lam, CW, James, JT, McCluskey, R, Hunter, RL. (2004) Pulmonary Toxicity of Single-Wall Carbon Nanotubes in Mice 7 and 90 Days After Intratracheal Instillation. Toxicol. Sci., 77, 126.

Léonard F, Talin AA. (2006) Size-dependent effects on electrical contacts to nanotubes and nanowires. Phys Rev Lett. 97(2):026804.

Li J, Andrews RJ (2007) Trimodal nanoelectrode array for precise deep brain stimulation: prospects of a new technology based on carbon nanofiber arrays. Acta Neurochir. Suppl. 97(Pt 2): 537–45.

Llinás RR, Walton KD, Nakao M, Hunter I, Anquetil PA (2005) Neuro-vascular central nervous recording/stimulating system: Using nanotechnology probes. J. Nanoparticle Res. 7:111–27.

Liopo AV, Stewart MP, Hudson J, Tour JM, Pappas TC (2006) Biocompatibility of native functionalized single-walled carbon nanotubes for neuronal interface. J. Nanosci. Nanotechnol. 6:1365–74.

Liu Z, Davis C, Cai W, He L, Chen X, Dai H (2008) Circulation and long-term fate of functionalized, biocompatible single-walled carbon nanotubes in mice probed by Raman spectroscopy. Proc. Natl. Acad. Sci. U S A 105:1410–1415.

Lovat V, Pantarotto D, Lagostena L, Cacciari B, Grandolfo M, Righi M, Spalluto G, Prato M, Ballerini L (2005) Carbon nanotube substrates boost neuronal electrical signaling. Nano Lett. 5:1107–10.

Lutolf MP, Hubbell JA (2005) Synthetic biomaterials as instructive extracellular microenvironments for morphogenesis in tissue engineering. Nat Biotechnol 23:47–55.

Magrez A, Kasas S, Salicio V, Pasquier N, Seo JW, Celio M, Catsicas S, Schwaller B, Forro L. (2006) Cellular Toxicity of Carbon-Based Nanomaterials. Nano Lett. 6:1121–1125.

Male KB, Hrapovic S, Luong JH (2007) Electrochemically-assisted deposition of oxidases on platinum nanoparticle/multi-walled carbon nanotube-modified electrodes. Analyst 132(12): 1254–61.

Markram H (2006) The blue brain project. Nature Rev. Neurosci. 7(2):153–60.

Marom S, Shahaf G (2002) Development, learning and memory in large random networks of cortical neurons: lessons beyond anatomy. Q. Rev. Biophys. 35(1):63–87.

Martel R, Derycke V, Lavoie C, Appenzeller J, Chan KK, Tersoff J, Avouris P. (2001) Ambipolar electrical transport in semiconducting single-wall carbon nanotubes. Phys. Rev. Lett. 87(25):256805.

Martinoia S, Massobrio P, Bove M, Massobrio G (2004) Cultured neurons coupled to microelectrode arrays: circuit models, simulations and experimental data. IEEE Trans. Biomed. Eng. 51(5):859–64.

Matsumoto, K, Sato, C, Naka, Y, Kitazawa, A, Whitby, RLD, Shimizu, NJ. (2007) Neurite outgrowths of neurons with neurotrophin-coated carbon nanotubes Biosci. Bioeng. 103, 216–220.

Mattson MP, Haddon RC, Rao AM (2000) Molecular functionalization of carbon nanotubes use as substrates for neuronal growth. J. Mol. Neurosci. 14:175–82.

Mazzatenta A, Giugliano M, Campidelli S, Gambazzi L, Businaro L, Markram H, Prato M, Ballerini L (2007) Interfacing neurons with carbon nanotubes: electrical signal transfer and synaptic stimulation in cultured brain circuits, J. Neurosci. 27(26):6931–6.

Migliore M, Cannia C, Lytton WW, Markram H, Hines ML (2006) Parallel network simulations with NEURON. J. Comput. Neurosci. 21(2):119–29.

Monthioux M, Kuznetsov VL (2006) Who should be given the credit for the discovery of carbon nanotubes? Carbon 44(9):1621–3.

Muguruma H, Shibayama Y, Matsui Y (2008) An amperometric biosensor based on a composite of single-walled carbon nanotubes, plasma-polymerized thin film, and an enzyme. Biosens. Bioelectron. 23(6):827–32.

Moxon KA, Kalkhoran NM, Markert M, Sambito MA, McKenzie JL, Webster JT (2004) Nanostructured surface modification of ceramic-based microelectrodes to enhance biocompatibility for a direct brain-machine interface. IEEE Trans. Biomed. Eng. 51(6):881–9.

Pappas TC, Wickramanyake WMS, Jan E, Motamedi M, Brodwick M, Kotov NA. (2007) Nanoscale Engineering of a Cellular Interface with Semiconductor Nanoparticle Films for Photoelectric Stimulation of Neurons. Nano Lett. 7:513–9.

Patolsky F, Weizmann Y, Willner I (2004) Long-Range Electrical Contacting of Redox Enzymes by SWCNT Connectors. Angew. Chem. Int. Ed. 43:2113–7.

Patolsky F, Timko BP, Yu G, Fang Y, Greytak AB, Zheng G, Lieber CM (2006) Detection, stimulation, inhibition of neuronal signals with high-density nanowire transistor arrays. Science 313:1100–4.

Phely-Bobin TS, Tiano T, Farrell B, Fooksa R, Robblee L, Edell DJ, Czerw R (2006) Carbon Nanotube Based Electrodes for Neuroprosthetic Applications. Mat. Res. Soc. Proc., San Francisco Spring Meeting.

Poland CA, Duffin R, Kinloch I, Maynard A, Wallace WAH, Seaton A, Stone V, Brown S, MacNee W, Donaldson K (2008) Carbon nanotubes introduced into the abdominal cavity of mice show asbestos-like pathogenicity in a pilot study. Nature Nanotech. 3:423–8.

Radosavljevic M, Appenzeller, J, Avouris P (2001) Engineering Carbon Nanotubes and Nanotube Circuits Using Electrical Breakdown, Science 292:706.

Rall W (1962) Electrophysiology of a dendritic neuron model. Biophys. J. 2:145–67.

Razal JM, Gilmore KJ, Wallace GG (2008) Carbon Nanotube Biofiber Formation in a Polymer-Free Coagulation Bath. Adv. Funct. Mater 18:61–6.

Robinson DA (1968) The Electrical Properties of Metal Microelectrodes, Proc. IEEE 56(6):1065–71.

Roy S, Vedala H, Roy AD, Kim DH, Doud M, Mathee K, Shin HK, Shimamoto N, Prasad V, Choi W (2008) Direct electrical measurements on single-molecule genomic DNA using single-walled carbon nanotubes. Nano Lett. 8(1):26–30.

Rutten WL (2002) Selective electrical interfaces with the nervous system. Annu. Rev. Biomed. Eng. 4:407–52.

Saito R, Fujita M, Dresselhaus G, Dresselhaus MS. (1992) Electronic structure of graphene tubules based on C60. Phys. Rev. B Condens. Matter 46(3):1804–11.

Sato Y, Shibata KI, Kataoka H, Ogino SI, Bunshi F, Yokoyama A, Tamura K, Akasaka T, Uo M, Motomiya K, Jeyadevan B, Hatakeyama R, Watari F, Tohji K (2005) Strict preparation and evaluation of water-soluble hat-stacked carbon nanofibers for biomedical application and their high biocompatibility: influence of nanofiber-surface functional groups on cytotoxicity. Mol. BioSyst. 1:142–8.

Sayes CM, Liang F, Hudson JL, Mendez J, Guo W, Beach JM, Moore VC, Doyle CD, West JL, Billups WE, Ausman KD, Vicki L. Colvin (2006) Functionalization density dependence of single-walled carbon nanotubes cytotoxicity in vitro, Toxicol. Lett. 161:135–42.

Schaefer AT, Larkum ME, Sakmann B, Roth A (2003) Coincidence detection in pyramidal neurons is tuned by their dendritic branching pattern. J Neurophysiol. 89(6):3143–54.

Schoen I, Fromherz P (2007) The mechanism of extracellular stimulation of nerve cells on an electrolyte-oxide-semiconductor capacitor. Biophys J. 92(3):1096–111.

Schwartz AB, Cui XT, Weber DJ, Moran DW (2006) Brain-controlled interfaces: movement restoration with neural prosthetics. Neuron 52(1):205–20.

Silva GA (2006) Neuroscience nanotechnology: progress, opportunities and challenges. Nature Rev. Neurosci. 7:65–74.

Sinha N, Ma J, Yeow JT (2006) Carbon nanotube-based sensors. J. Nanosci. Nanotechnol. 6(3):573–90.

Sirivisoot S, Yao C, Xiao X, Sheldon BW, Webster TJ (2007) Greater osteoblast functions on multiwalled carbon nanotubes grown from anodized nanotubular titaniu for orthopedic applications. Nanotechnology 18:365102.

Snow ES, Novak JP, Campbell PM, Park D (2003) Random networks of carbon nanotubes as an electronic material, Appl. Phys. Lett. 82:2145.

Storace M, Bove M, Grattarola M, Parodi M (1997) Simulations of the behavior of synaptically driven neurons via time-invariant circuit models, IEEE Trans. Biomed. Eng. 44(12):1282–7.

Sucapane A, Cellot G, Prato M, Giugliano M, Parpura V, Ballerini L (2008) Interactions between cultured neurons and carbon nanotubes: A nanoneuroscience vignette. J. Nanoneurosci. 1:1–7.

Tans SJ, Verschueren ARM, Dekker C (1998) Room-temperature transistor based on a single carbon nanotube Nature 393:49–52.
Tasis D, Tagmatarchis N, Bianco A, Prato M (2006) Chemistry of carbon nanotubes. Chem. Rev. 106:1105–36.
Traub RD, Miles R (1991) Neuronal Networks of the Hippocampus, Cambridge, UK: Cambridge University Press.
Tsang SC, Guo Z, Chen YK, Green MLH, Hill HAO, Hambley TW, Sadler PJ (1997) Immobilization of Platinated and Iodinated Oligonucleotides on Carbon Nanotubes. Angew Chem. Int. Ed. 36:2198–2200.
Yao Z, Kane CL, Dekker C. (2000) High-field electrical transport in single-wall carbon nanotubes. Phys. Rev. Lett. 84(13):2941–4.
Wang K, Fishman HA, Dai H, Harris JS (2006) Neural Stimulation with a Carbon Nanotube Microelectrode Array, Nano Lett. 6:2043–8.
Wörle-Knirsch JM, Pulskamp K, Krug HF (2006) Oops They Did It Again! Carbon Nanotubes Hoax Scientists in Viability Assays, Nano Lett. 6(6):1261–8.
Zhang M, Liu K, Xiang L, Lin Y, Su L, Mao L (2007) Carbon Nanotube-Modified Carbon Fiber Microelectrodes for In Vivo Voltammetric Measurement of Ascorbic Acid in Rat Brain. Anal. Chem. 79(17):6559–65.
Zanello LP, Zhao B, Hu H, Haddon RC (2006) Bone Cell Proliferation on Carbon Nanotubes, Nano Lett. 6(3):562–7.

# Carbon Nanotubes as Modulators of Neuronal Growth

**Reno C. Reyes and Vladimir Parpura**

## 1 Introduction

Carbon nanotubes (CNTs) have emerged as one of the most promising nanomaterials that can be used in neurosciences. Recently, the use of CNTs for modulation of neural cell growth has been an area of active research. CNTs, bare or functionalized with various chemical groups and biologically relevant molecules, are compatible with neuronal cell growth. Rather than reporting on the structure, properties and functionalization of CNTs, that we have previously reviewed elsewhere [for detailed description, see (Bekyarova et al. 2005a; Bekyarova et al. 2005b)] in this chapter, we focus our discussion on a subset of biological applications of both single-walled (SW) and multi-walled (MW) CNTs: (1) as scaffolds/substrates for adhesion and growth of neurons and (2) as water-dispersible agents to modulate neurite outgrowth. We begin by presenting the evidence that bare CNTs can be used as permissive scaffolds. We, then, discuss the use of functionalized CNTs, which can modulate neuronal growth in a graded manner; positively charged CNT scaffolds permitted neurite outgrowth of neurons in culture to a greater extent than when these cells were grown on zwitterionic or negatively charged CNTs. Additional physical properties, conductivity

R.C. Reyes
Department of Neurobiology, Center for Glial Biology in Medicine,
Atomic Force Microscopy and Nanotechnology Laboratories,
Civitan International Research Center, Evelyn F. McKnight Brain Institute, University of Alabama, Birmingham, AL, USA

Department of Neurology, University of California, San Francisco and Veterans Affairs Medical Center, San Francisco, CA, USA

V. Parpura (✉)
Department of Neurobiology, Center for Glial Biology in Medicine, Atomic Force Microscopy and Nanotechnology Laboratories, Civitan International Research Center, Evelyn F. McKnight Brain Institute, University of Alabama, Birmingham, AL, USA
e-mail: vlad@uab.edu

G.A. Silva and V. Parpura (eds.), *Nanotechnology for Biology and Medicine: At the Building Block Level*, Fundamental Biomedical Technologies,
DOI 10.1007/978-0-387-31296-5_10, 

and roughness, of CNT scaffolds can also affect neuronal morphology. In the follow-up section, we present the use of water-dispersible CNTs as agents that can be applied to neurons and affect neurite outgrowth. It is this approach that has been demonstrated to aid regeneration after spinal cord injury in vivo. We only discuss the use of neurons in primary culture, rather than reporting on results using various cell lines established from neural cells, which we have reviewed elsewhere (Lee and Parpura 2009). When information is available, we also report on the brain regions from which primary neurons were isolated and on the animal type/species used, since some of the reported effects could be species and region specific. An emerging picture is that CNTs are neuron-compatible substrates/scaffolds and injectable agents which may find future utilization in medicine.

## 2 CNTs as Retainable Substrates/Scaffolds for Neuronal Growth

The first evidence for applicability of CNTs as a permissive planar substrate for neuronal growth came from the study by Mattson et al. (Mattson et al. 2000). By using fixed materials and scanning electron microscopy (SEM), the authors demonstrated that hippocampal neurons from primary cultures of prenatal (embryonic day 18) rat brains survived for at least 8 days, and formed one or two neuronal processes, neurites, when grown on the surfaces of bare/unmodified MWCNTs. Such MWCNTs were prepared by a catalytic decomposition of ferrocene–xylene mixture (Andrews et al. 1999) resulting in MWCNT sheets containing MWCNTs with diameters of ~20 nm and lengths between 20–200 μm. MWCNT sheets were dispersed in ethanol by sonication. Dispersed MWCNTs were then deposited onto glass coverslips precoated with polyethyleneimine (PEI), a cationic polymer commonly used to grow neural cells in culture. After ethanol evaporation, the resulting glass coverslip surface contained patches of MWCNTs on top of the PEI layer as well as areas coated with PEI alone so that neurons could grow on either/both of these surfaces. The direction of the neurite outgrowth from somata of cultured hippocampal neurons was not influenced by the orientation of the MWCNTs as neurites grew across or along CNTs. Additionally, neurons displayed more preponderant and branched neurites when they were grown on top of the PEI surface, rather than on top of MWCNTs. These observations suggest that although unmodified MWCNTs are a permissive substrate for neuronal growth and neurite outgrowth, they may restrict some elements of neuronal growth. Thus, if enhanced neurite outgrowth and multibranching are of interest, then MWCNTs alone may not be an optimal cell scaffold/substrate and their modifications should be considered. To address whether functionalization of MWCNTs with biological molecules can be used to control the number of neurites, their outgrowth, and branching, Mattson et al. noncovalently coated MWCNTs with 4-hydroxynonenal (4-HNE) by physical adsorption. The rationale for the usage of this lipid peroxidation product came from its effect on intracellular $Ca^{2+}$ levels in cultured hippocampal neurons (Mark et al. 1997). Since $Ca^{2+}$ influx can regulate neurite elongation (Kater et al. 1988), it is likely that 4-HNE would also exhibit such action. Indeed, neurons grown on MWCNTs coated with 4-HNE exhibited increased neurite number, length, and branching when compared to neurons grown on bare MWCNTs. This effect appeared to be mediated by 4-HNE, since increased growth was not evident when neurons were grown on MWCNTs in which adsorption of 4-HNE was carried out in an excess of histidine that binds to 4-HNE (Uchida and Stadtman 1992). Taken together, this study was the first to show that both bare and functionalized CNTs are biocompatible with neuronal survival and growth. Furthermore, noncovalent modifications of CNTs with a biomolecule can be used to modulate neurite outgrowth and neurite branching.

Coating CNTs with biological molecules using adsorption may not be a viable option for generation of long-term scaffolds. Namely, while MWCNTs are inert materials and are readily available for long-term exposure to neural tissue/cells, biological compounds could degrade. Furthermore, a biologically active compound applied by physical adsorption could simply dissociate and undergo diffusion, thus effectively lowering its local concentration. Consequently, a more permanent modification of CNTs by covalent linking of various stable chemical groups and molecules would be a desirable alternative functionalization strategy. Hu et al. (2004) provided evidence that neuronal growth can be modulated by covalent modifications made predominantly at the ends of MWCNTs to systematically vary their surface charge. At the onset of the study, Hu et al. used the so called as-prepared MWCNTs (AP-MWCNTs) that were produced by chemical vapor deposition (CVD). These AP-MWCNTs were deposited onto glass coverslips precoated with PEI and used as scaffolds/substrates for neuronal growth. Initially, neuronal cultures, prepared from hippocampi of 0- to 2-day-old Sprague-Dawley rats, were fixed and examined using SEM. As expected, both PEI and AP-MWCNTs were permissive substrates for neuronal growth, albeit neurite outgrowth was parsimonious on MWCNTs when compared to that of neurons grown on PEI (Fig. 1a and b). Neurons on CNTs grew at least 1 week in culture. These findings are consistent with previous work utilizing fixed neuronal cultures (Mattson et al. 2000). Similar data was obtained when living neurons were studied using the vital stain calcein and fluorescence microscopy (Fig. 1c and d). Again, neurons grown on positively charged PEI outperformed neurons grown on presumably neutrally charged AP-MWCNTs. However, since PEI and AP-MWCNTs also assume different topology, any comparison in regard to the sole role of charge on neuronal growth could not be made. To further investigate whether surface charge could be exploited as a modulator of neuronal growth and neurite outgrowth, various functionalizations of the MWCNT backbone were prepared. Here, AP-MWCNTs were refluxed in nitric acid to remove the metal catalyst residues; this procedure additionally led to uncapping of CNTs and termination of their open ends with carboxyl groups resulting in functionalized MWCNT-COOH. Further reaction of MWCNT-COOH with oxalyl chloride resulted in the acyl chloride intermediate MWCNT-COCl, which was then used for covalent attachment of either ethylenediamine (EN) or poly-*m*-aminobenzene-sulfonic acid (PABS) to produce two additional final products: MWCNT-EN and MWCNT-PABS. Thus, using this method, Hu et al. generated MWCNTs with

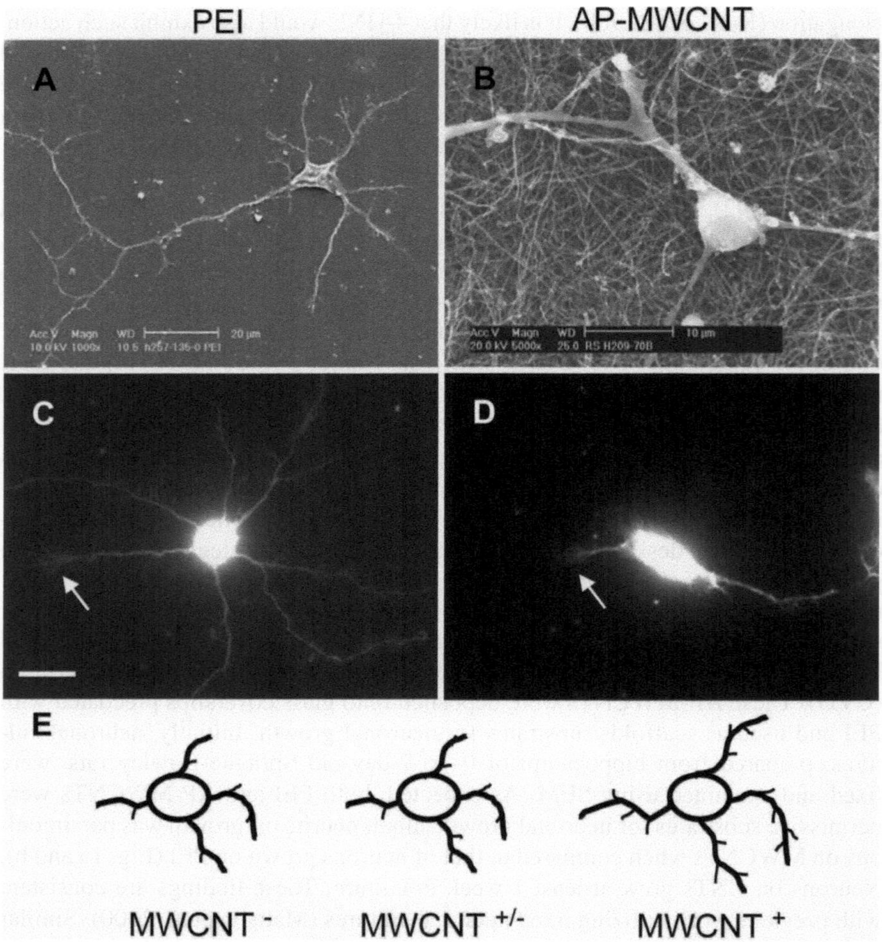


**Fig. 1** Bare MWCNTs are a permissive substrate for neuronal growth. Both polyethyleneimine (PEI; *left column images*) and as-prepared MWCNTs (AP-MWCNTs; *right column images*) allow neuronal viability and neurite outgrowth. (**a, b**) SEM images of hippocampal neurons grown on PEI (**a**) and AP-MWCNTs (**b**). (**c, d**) Fluorescence images showing live neurons, which accumulate a vital stain, calcein. Arrows indicate growth cones. Scale bar: 20 μm, except 10 μm in **b**. (**e**) Drawing summarizing the effects of MWCNT charges on the neurite outgrowth and branching. Neurons grown on positively charged MWCNTs have longer and more branched neurites than neurons grown on negative or zwitterionic MWCNTs. Modified from (Hu et al. 2004)

comparable typology but, due to dissociation constants of attached groups, with different charges, negatively (MWCNT-COOH), zwitterionic (MWCNT-PABS), or positively charged (MWCNT-EN) at the physiological extracellular pH of 7.35. Indeed, the authors used mid-infrared (IR) spectroscopy of MWCNTs to confirm the success of covalent modifications. Each of these three functionalized MWCNTs was deposited onto glass coverslips precoated with PEI and used as scaffolds/

substrates for neuronal growth. Fixed neurons were investigated using SEM while live cells loaded with calcein were subjected to fluorescence microscopy. The morphological characteristics of neurons grown on zwitterionic MWCNT-PABS were similar to those of neurons grown on AP-MWCNT. Neurite outgrowth was dependent on the surface charge carried by functionalized MWCNTs. Whereas there was no significant difference in the number of neurites between neurons grown on the various MWCNTs, neurons grown on the positively charged MWCNT-EN and the zwitterionic MWCNT-PABS had significantly greater number of growth cones than the neurons grown on the negatively charged MWCNT-COOH. In addition, neurons grown on positively charged MWCNT-EN produced significantly longer neurites and greater neurite branching than neurons grown on zwitterionic or negatively charged MWCNTs. Neurite branching was graded, since neurons grown on zwitterionic MWCNT-PABS showed less branches than neurons grown on MWCNT-EN, but more than neurons grown on negatively charged MWCNT-COOH (Fig. 1e). Taken together, the results of this study suggest that MWCNT modifications provide a means to modulate the growth of neurons based on electrostatic charge that can be introduced to MWCNTs by the attachment of various chemical compounds.

Several additional research groups made use of MWCNTs as a substrate/scaffold for growth on neurons in primary culture. In a technical note, Xie et al. (2006) implicated that functionalized MWCNTs supported greater neurite growth than nonfunctionalized MWCNTs. In this study, MWCNTs were obtained by microwave CVD (Varadan and Xie 2002). After refluxing MWCNTs in the mixture of sulfuric and nitric acids, an oxidation method that introduces carboxyl groups mainly at the end of MWCNTs (Liu et al. 1998), the resulting MWCNT-COOH material was dispersed in water and then vacuum filtered to generate MWCNT mats on top of a track-etched polycarbonate membrane. These mats were seeded with dissociated dorsal root ganglion (DRG) neurons from newborn rats. Although the pH of the standard culture medium used (DMEM) was not explicitly stated, it should be ~7.4, which implicates that MWCNT-COOH mats display negative surface charge as opposed to neutrally charged unmodified/bare MWCNT mats. Using SEM, Xie et al. qualitatively assessed neuronal growth on bare and functionalized MWCNTs. They observed growth cone formation and neurites, which branched and some also bundled, extending from DRG neurons grown on MWCNT-COOH mats. DRG neurons cultured on bare mats showed similar morphological characteristics, but shorter neurites were observed. Furthermore, neurites readily intertwined with functionalized, but not with bare MWCNTs, implicating that there might be preferential/stronger neurite interactions with functionalized rather than with bare MWCNTs. These qualitative observations in respect to neurite outgrowth using DRG neurons are in disagreement with the quantitative analysis done on live hippocampal neurons as described by Hu et al. (Hu et al. 2004). Such seemingly disparate findings could be a result of: (1) differences in the experimental approach: qualitative SEM on fixed (Xie et al. 2006), as opposed to quantitative fluorescence microscopy of live neurons (Hu et al. 2004); (2) reporting on variability in growth by neurons isolated from different regions: peripheral (DRG) vs. central (hippocampus) nervous system. Nonetheless, these differences should not detract from the fact that (un) functionalized CNTs can serve as permissive substrates/scaffolds for neuronal growth and that the neurite outgrowth can be influenced by scaffold properties.

Deposition of AP-MWCNTs to glass coverslips can be problematic, so it may result in detachment of MWCNT mats which then float to the surface of the culture media (Lovat et al. 2005). To improve on MWCNT substrate adhesion to glass coverslips, Lovat et al. (Lovat et al. 2005) developed an alternative MWCNT deposition procedure. First, they subjected AP-MWCNTs to 1,3-dipolar cycloaddition of azomethine ylides, which generates pyrrolidine (PYR) groups on the MWCNT ends and sidewalls. This reaction confers MWCNT-PYR with increased solubility in organic solvents. Hence, the MWCNT-PYR material was dissolved in dimethylformamide and applied to glass coverslips. Following evaporation of the solvent, coverslips containing MWCNT-PYR were baked to remove PYR groups. The resulting glass coverslips contained a retainable film of bare/defunctionalized MWCNTs, which was used as a substrate for growth of hippocampal neurons. Since morphological parameters of neuronal growth were not the subject of this study, the comparison of growth and neurite outgrowth on this substrate to those described above using various scaffolds is not possible. However, the neurons grown on top of MWCNT films survived and formed functional networks displaying spontaneous synaptic transmission within 8–10 days in culture. The detailed description of this approach and the effects that interactions between neurons and MWCNTs can have on synaptic transmission is available elsewhere [Chap. 9 in this volume; also see (Lovat et al. 2005; Mazzatenta et al. 2007; Cellot et al. 2009)].

Thus far, we only discussed the usage of MWCNTs as planar substrates to support neuronal growth. Galvan-Garcia et al.(Galvan-Garcia et al. 2007) presented evidence that nonfunctionalized MWCNTs not only in the form of sheets, but also as yarns, can serve as permissive substrates/scaffolds for neuronal attachment and growth. They generated MWCNT sheets using the catalytic CVD. MWCNT yarns were made by a subsequent spinning of these sheets. MWCNTs are directionally oriented in both forms. Galvan-Garcia et al. evaluated permissiveness of bare MWCNT sheets and yarns as a substrate for neuronal growth in culture. Both materials were biocompatible since they promoted cell attachment, differentiation, growth, and long-term survival of various cell types, including dissociated cortical and cerebellar neurons from rat pups (postnatal day 0–10) and neonatal mice DRG explants containing neurons. Using bright and fluorescence light microscopy, they demonstrated that DRG neurons survived over 2 weeks in culture. These neurons extended their axons from explants onto MWCNT sheets; their neurites could attach and extend along MWCNT yarns. Dissociated cortical and cerebellar neurons displayed neurite outgrowth to a similar extent as when they were seeded onto MWCNT sheets or onto the commonly used polycationic substrate for neural cell growth, polyornithine (PO), which was precoated on glass coverslips. Interestingly, cortical and cerebellar neurons cultured on MWCNT sheets showed larger growth cones than when grown on PO while the length of neurites was not significantly different. Based on electrostatic charge alone, this is an unexpected finding since MWCNT sheets should be neutrally charged in DMEM while PO should be positively charged. However, besides the charge difference between PO and MWCNT, these substrates highly likely also differed in their surface topology. In contrast, in the Hu et al. study (Hu et al. 2004), the same MWCNT topology was used to display different charges. Thus, these data may point to the possibility that, besides electrostatic charge, other physical qualities of the substrate could affect neuronal growth, a topic that we further discuss below.

As alluded to above, substrate qualities play a role in the process of growth cone motility and neurite branching (Lustgarten et al. 1991; Mattson et al. 2000). First, we discuss "dilution" of electrostatic charge and then the effects of roughness/topology and conductivity on neuronal growth. Since the "flavor" (negative, zwitterionic/neutral, and positive) of electrostatic charge of substrate matters, it is possible that reducing the charge of the substrate could be used to further modulate neuronal growth and neurite outgrowth. One approach is to reduce PEI cationic charge with bare CNTs that are neutral. Hu et al. (Hu et al. 2005) did just that by synthesizing a polymer, where branched PEI was grafted onto SWCNTs. In this study, commercially available SWCNT-COOH material was treated with oxalyl chloride to form the intermediate product SWCNT-COCl. In turn, branched PEI was reacted with SWCNT-COCl to generate the graft copolymer SWCNT-PEI. Quantitative assessment of the SWCNT percent weight in the graft copolymer using thermogravimetric analysis and near-IR spectroscopy revealed that SWCNTs contribute 18–19% of SWCNT-PEI weight. When deposited onto glass coverslips and visualized under bright-field microscopy, as expected, this material showed intermediate transparency to that of PEI and AP-MWCNT substrates (Fig. 2a–c). Comparison of morphological characteristics of calcein-loaded hippocampal neurons in cultures (Fig. 2d–f), prepared from 0- to 2-day-old rats, revealed that neurons grown on SWCNT-PEI display intermediate growth and neurite outgrowth characteristics to those of neurons grown on AP-MWCNT- or PEI-coated coverslips (Fig. 2g). Thus, it appears that the intermediate growth of neurons on SWCNT-PEI can be attributed to the reduced positive charge of the PEI as it is copolymerized with SWCNTs. Whether graded dilution of charge would yield graded neuronal growth and neurite outgrowth awaits further experimentation.

Roughness of the substrate can affect neuronal growth and interactions between neurons and the substrate. For instance, in a non-CNT system with $SiO_2$ as the substrate, it was demonstrated that roughness of the material dictated growth of substantia nigra neurons cultured after isolation from prenatal Wistar rats (Fan et al. 2002). It was reported that after 5 days in culture about as twice as many neurons per area adhered to and grew on a $SiO_2$ surface having an average roughness (Ra) of 20–50 nm, when compared to $SiO_2$ surfaces with their Ra less than 10 nm or greater than 70 nm. A later study employing CNTs as scaffolds for neural cells had made similar observations that topology/roughness of the substrate matter. Hence, Sorkin et al. (2009) cultured dissociated neurons that originated from embryonic rat cortices or from locust frontal ganglion on quartz micropatterned with AP-CNT (most likely MWCNTs) islands made by CVD. Neurons grown on patterned substrates were visualized after fixation using SEM. While neurons from either species were unable to adhere to $SiO_2$, they grew on CNT islands. Although not systematically studied, this effect was attributed to surface roughness of CNT islands. Rat neurons,

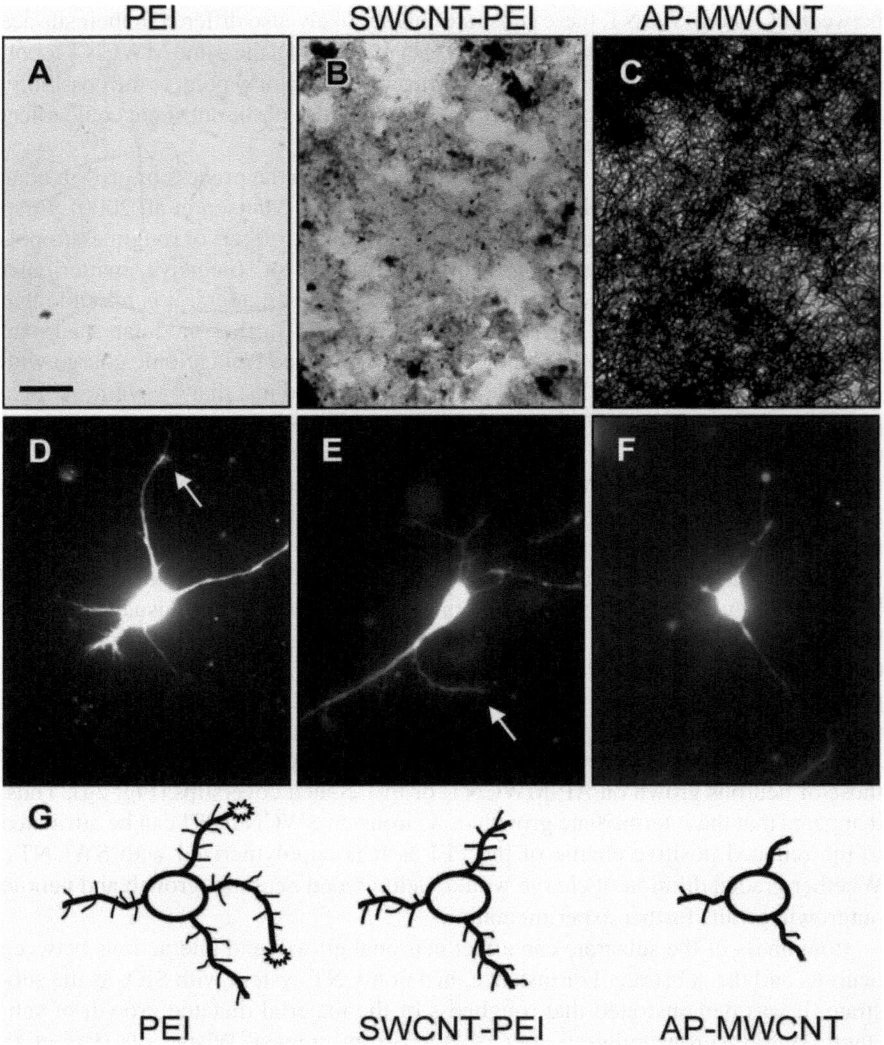


**Fig. 2** Bright-field light microscopy images of glass coverslips coated with PEI (**a**), SWCNT-PEI (**b**), and AP-MWCNTs (**c**). Fluorescence images show live hippocampal neurons accumulating calcein and grown on PEI (**d**), SWCNT-PEI (**e**), and AP-MWCNT (**f**). Arrows indicate growth cones. Scale bar, 20 μm. (**g**) Drawing summarizing the effect of charge "dilution" on the neurite outgrowth and branching. Neurons grown on SWCNT-PEI graft copolymer display neurite outgrowth and branching intermediate to those exhibited when grown on PEI and AP-MWCNTs. Modified from (Hu et al. 2005)

which have thinner processes than locust neurons, showed extensive curling and intertwining of neurites around CNTs. A similar phenotype was seen with thin, but not thick, neurites of locust neurons; thick neurites show entanglement between themselves rather than with CNTs. It appears that CNTs with diameters relatively similar to that of thin neurites allowed entanglement, which could represent an anchoring mechanism for neurite attachment to rough surfaces. These results suggest that surface topology of CNT scaffolds should be considered when designing scaffolds for neuronal growth.

Conductivity is another important CNT attribute which can modulate neuronal growth and neurite outgrowth in cell culture. Malarkey et al. (Malarkey et al. 2009) seeded hippocampal neurons, prepared from 0- to 2-day-old rats, onto retainable CNT films deposited on glass coverslips. Such films were made by spraying water-soluble SWCNTs on heated glass coverslips. Water-soluble SWCNTs were obtained by functionalization of SWCNTs with polyethylene glycol (PEG). Here, the SWCNT-COCl intermediate was reacted with PEG to generate SWCNT-PEG. Malarkey et al. sprayed different amounts of the SWCNT-PEG aqueous solution to generate retainable uncharged substrates/films with the thickness of 10, 30, or 60 nm (Fig. 3a), which corresponded to the increasing film conductivity of 0.3 S/cm, 28 S/cm, and 42 S/cm, respectively. While the conductivities increased with increasing thickness, the surface roughness of these films was not significantly different from one another as characterized by atomic force microscopy. However, their surface was significantly rougher that that of the standard, positively charged, and nonconductive substrate PEI. Thus, effects on neuronal growth achieved on the various SWCNT-PEG films result from their differences in conductivity, not surface roughness or charge. It should be noted that when the neuronal growth on various SWCNT films is compared to that of neurons grown on PEI, any difference observed could be the outcome of roughness, charge, and/or conductivity. To assess possible effects on cellular morphology, hippocampal neurons were cultured on PEI and SWCNT-PEG films for 3 days. At that time, live neurons were loaded with calcein and visualized using fluorescence microscopy (Fig. 3b–e). The total number of neurites decorating individual neurons remained the same regardless of the conductivity of the film. However, the total length of all processes and their branches, as well as the mean neurite length, was significantly greater in neurons grown on the 10-nm thick SWCNT-PEG films than in neurons grown on coverslips coated with the PEI or with the thicker films (30 and 60 nm) with higher conductivity, but comparable roughness to the 10-nm SWCNT-PEG film (Fig. 3f). These findings indicate that certain SWCNT-PEG conductivity (0.3 S/cm) could promote neurite outgrowth as compared to other SWCNT films. Additionally, there was a significant increase in the average area of the neuronal cell body at 28 S/cm when compared to standard PEI and to 42 S/cm substrate, but not when compared to 0.3 S/cm substrate, on which neurons showed only a trend of body enlargement (Fig. 3f). There was a significantly higher number of growth cones on neurons grown on 60-nm SWCNT films when compared to measurements on 30-nm SWCNT films. Since the roughness of these two conductive films is similar, it appears that the higher conductivity caused an increase in the number of growth cones. Neurons grown on the smoother

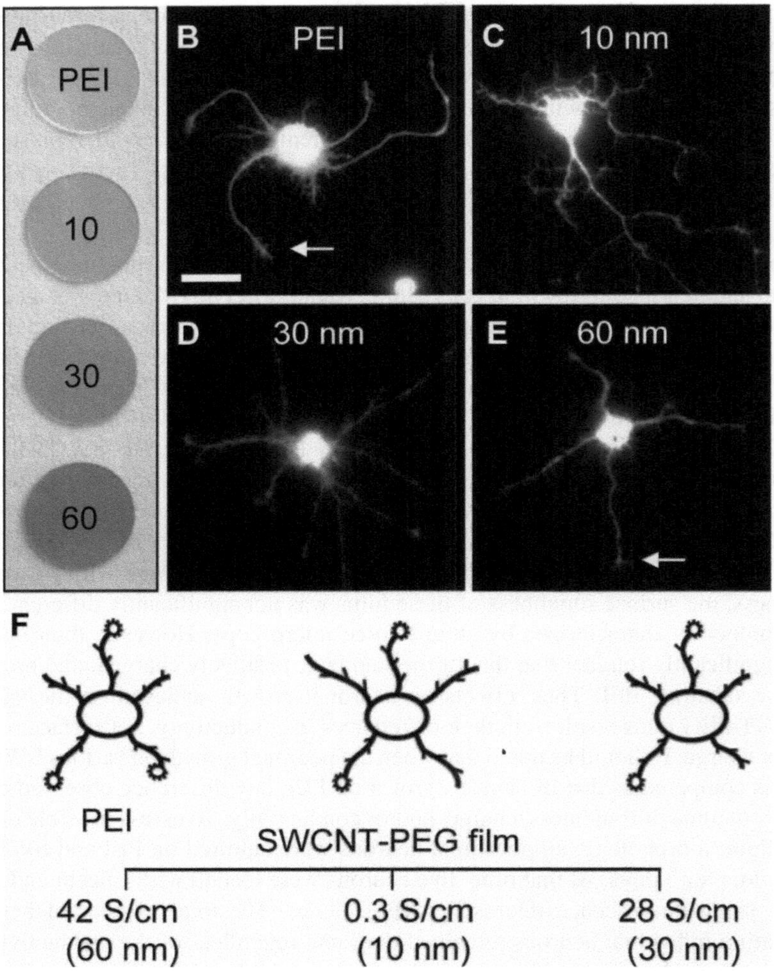


**Fig. 3** Image of 12 mm in diameter glass coverslips (**a**) coated, from top to bottom, with: PEI (standard), 10-, 30-, and 60-nm-thick SWCNT-PEG films. Fluorescence images of live hippocampal neurons grown on coverslips coated with PEI coated (**b**) or SWCNT-PEG films of varying thickness/conductivity (**c**, 10-nm thick; **d**, 30-nm thick; **e**, 60-nm thick). Arrows indicate growth cones. (**b–e**) Scale bar, 20 μm. (**f**) Drawing summarizing the effects of the conductivity of SWCNT substrates on neurite outgrowth, growth cones, and body size. The neurite outgrowth was significantly greater in neurons grown on the 10-nm thick SWCNT-PEG films. The average area of the neuron cell body grown on the 30-nm thick SWCNT films was enlarged. Neurons grown on the PEI had a significantly higher number of growth cones than those grown on 10- and 30-nm SWCNT films, but not higher than those grown on 60-nm SWCNT films. Conductivity of each film is listed in S/cm; PEI is not conductive. Modified from (Malarkey et al. 2009)

PEI had a significantly higher number of growth cones than those grown on 10- and 30-nm SWCNT films, but not higher than those grown on 60-nm SWCNT films (Fig. 3f). Since positively charged PEI has a much smoother surface than SWCNT films, it appears that perhaps a combination of charge, roughness, and/or conductivity could cause this effect. Taken together, these results indicate that a SWCNT-PEG substrate can modulate neuronal growth and neurite outgrowth within a narrow range of its conductivity. The studies discussed above showed that various qualities of CNT substrates/scaffold, most notably their charge, roughness/topology, and conductivity, can modulate neuronal growth and neurite outgrowth. Consequently, these physical properties of CNTs should be taken into consideration when designing CNTs as scaffolds for neuronal growth.

## 3 Water-Dispersible CNTs as Modulators of Neuronal Growth

Experimental approaches discussed thus far aimed to explore the effects of CNTs as biocompatible substrate/scaffolds. Of course, an additional strategy is to generate CNTs that are soluble/dispersible in aqueous media of extracellular space of the brain so that they could be delivered as diffusive agents to affect neurite outgrowth. Ni et al. (Ni et al. 2005) addressed this issue by treating neurons grown on PEI substrate with water-soluble SWCNTs that were functionalized with either PEG or PABS via the SWCNT-COCl intermediate. At physiological pH, these graft copolymers are neutral or zwitterionic, respectively. Using calcein-loaded hippocampal neurons, prepared from 0- to 2-day-old rats, and fluorescence microscopy (Fig. 4a–c), they demonstrated that neurons treated with either form of soluble SWCNTs showed a reduced number of neurites and growth cones when compared to control (sham treated with the vehicle) neurons. Coincidentally, neurons treated with water-soluble SWCNTs also exhibited longer neurites than controls (Fig. 4d); these effects were concentration dependent. Ni et al. investigated a possible mechanism underlying actions of water-soluble SWCNTs on neurite outgrowth. It appears that SWCNT-PEG reduced $Ca^{2+}$ influx from the extracellular space through voltage-dependent $Ca^{2+}$ channels (VDCCs), known regulators of neurite elongation (Mattson and Kater 1987; Kater et al. 1988; Mattson et al. 1988). This was investigated using the intracellular calcium indicator fluo-3. When compared to control, neurons treated with SWCNT-PEG had reduced cytosolic $Ca^{2+}$ accumulation after depolarizing them with high extracellular potassium (50 mM) to open VDCCs; the depolarization-dependent $Ca^{2+}$ entry was sensitive to the VDCC blocker $Cd^{2+}$. An increase in neuronal intracellular $Ca^{2+}$ levels can regulate plasma membrane/vesicular recycling, which has been implicated to play a role in the rate of neurite elongation (Zakharenko and Popov 2000). Consequently, Malarkey at al. examined whether SWCNTs could affect membrane recycling (Malarkey et al. 2008). Reasoning that the fluorescent dye *N*-(3-triethylammoniumpropyl)-4-(4-(dibutylamino)styrl)pyridinium dibromide (FM 1–43) is plasma membrane impermeable and is taken up by cells through endocytosis, they compared the effect of SWCNT-PEG exposure on FM 1–43 load in

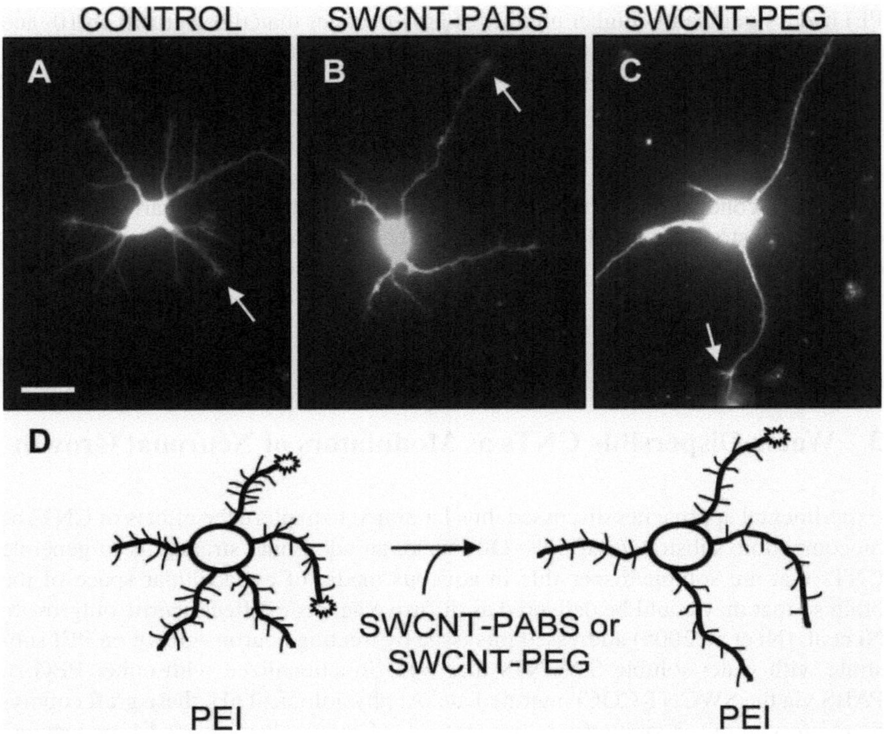


**Fig. 4** Chemically functionalized water-soluble SWCNTs added to the culture medium modulate neurite outgrowth. Fluorescence images of live hippocampal neurons, accumulating the vital stain, calcein. Neurons grown on PEI-coated glass coverslips (**a**; control, sham treated) can be treated with water-soluble CNTs (each at 1 μg/mL), either SWCNT-PABS (**b**) or SWCNT-PEG (**c**) to affect their growth characteristics. Arrows indicate growth cones. Scale bar, 20 μm. (**d**) Drawing summarizing the effects of water-soluble SWCNTs on neurite outgrowth and growth cones. Water-soluble SWCNT-PABS or SWCNT-PEG graft copolymers when added to the culturing medium of neurons grown on PEI substrate increased the length of selected neurites and reduced the number of growth cones. Modified from (Ni et al. 2005)

neurons and found no significant differences in the dye load in neurons at rest that were exposed to the different concentrations of SWCNT-PEG. This indicated that constitutive membrane recycling was not affected by SWCNT-PEG. However, data indicated that SWCNT-PEG inhibited depolarization-dependent load of the dye (Malarkey et al. 2008). Subsequent experiments indicated that the inhibitory action of SWCNT-PEG was preferentially affecting regulated endocytosis. Hence, the exocytotic incorporation of vesicles into the plasma membrane was not balanced by the endocytotic retrieval in the presence of SWCNTs. This could effectively cause the increase in neurite length observed by Ni et al. (Ni et al. 2005). It should be noted that both studies (Ni et al. 2005; Malarkey et al. 2008) were done by the same group under the same experimental conditions. The reduction of depolarization-dependent

$Ca^{2+}$ accumulation (Ni et al. 2005) and the inhibition of regulated endocytosis (Malarkey et al. 2008) both showed concentration dependence on SWCNTs, which corresponded well to SWCNT-PEG concentrations affecting neurite numbers and outgrowth (Ni et al. 2005). Taken together, these results suggest the exciting possibility that water-soluble SWCNTs could be delivered locally to the site of CNS injury to enhance neurite outgrowth which might increase the probability to overpass the site of injury and aid in the process of regeneration.

A possible therapeutic intervention using water-soluble SWCNTs has been initiated for treatment of spinal cord injury (SCI) (Roman et al. 2009a; Roman et al. 2009b; Roman et al. 2011). Traumatic SCI causes tissue damage resulting in the formation of a cavity that inhibits axonal regrowth. Filling this cavity with a growth-inducing agent, such as SWCNT-PEG, could promote regeneration and repair. To assess this issue, SCI was induced by complete transection of the spinal cord at the thoracic 9 vertebrae of adult female Sprague-Dawley rats. Immediately after transection, the epicenter of the lesion was injected with either SWCNT-PEG solution or the vehicle. At 28 days post SCI, the rats were euthanized and spinal cord tissue was examined using immunocytochemistry. The addition of SWCNT-PEG was found to promote neurite outgrowth and reduce reactive gliosis. These data suggest that injectable water-soluble SWCNT-PEG materials hold promise as a neurite outgrowth-promoting agent in treatment after SCI.

The above studies suggest that CNTs are an excellent source material for generating injectable materials that can be used to modulate neurite outgrowth in culture and in vivo. While mechanisms underlying such actions have been investigated, it appears that they are mediated simply by physical–chemical characteristics of functional groups and/or of CNTs. One caveat of such an approach is that it may lack specificity that neural cells utilize in intercellular interactions. Consequently, it should be important to engineer CNTs that are functionalized with biological molecules that possess ligand–receptor specificity. Matsumoto et al. (Matsumoto et al. 2007) have accomplished a first step toward such a goal. They functionalized CNTs using endogenous ligands in the CNS, neurotrophins, to assess the retention of ligand activity when conjugated to CNTs. Matsumoto et al. covalently bonded nerve growth factor (NGF) or brain-derived neurotrophic factor (BDNF) to MWCNTs, and studied the effect of these graft copolymers, delivered by dispersing them in culturing media of neurons isolated from DRGs of 8-day-old chick embryos. To generate neurotrophin-bearing MWCNTs, CNTs were first functionalized with amino groups that were then used to attach NGF or BDNF. The resulting neurotrophin-MWCNT materials were dispersed in phosphate-buffered saline by sonication and the success of this functionalization procedure was assessed using an enzyme-linked immunoabsorbent assay and antibodies against NGF or BDNF. DRGs from chick embryos were dissociated to establish neuronal cultures that were grown in the standard laminin-coated well plates. The numbers of DRG neurons with neurite outgrowth longer than the cell body were counted. The addition of a dispersion of either NGF-MWCNT or BDNF-MWCNT in cell culture media prompted neurite outgrowth, which was comparable to that caused by the soluble NGF or BDNF. This indicated that neurotrophins covalently attached to MWCNTs retained their bioactivity, including the engagement of extracellular signal-regulated kinase (ERK) pathway as determined in a subsequent study using NGF-MWCNTs (Matsumoto et al. 2010). Future experiments will have to be carefully designed to assess the possible use of CNTs functionalized with neurotrophins for support of neurite outgrowth in therapy after brain injury.

## 4 Concluding Remarks

The intent of this chapter is to discuss the use of CNTs in neurobiology as scaffolds and water-dispersible injectable compounds to affect neuronal growth and neurite outgrowth. From the presented body of work, there are palpable indices that these nanomaterials could find valuable application in medicine. Indeed, the exploitation of CNTs is only beginning. One can only start to imagine further applications of these materials in neurosciences based on their physical properties and the versatilities of available chemical modifications. Of course, as nanotechnology further advances, there should be various additional materials generated that together with CNTs could aid our ability to repair the loss of brain function due to injury and/or to better understand the operation of this organ.

**Acknowledgments** We thank Randy F. Stout, Jr. and William Lee for comments on this manuscript. This work was supported by the National Science Foundation (CBET 0943343).

## References

Andrews R, Jacques D, Rao AM, Derbyshire F, Qian D, Fan X, Dickey EC, Chen J (1999) Continuous production of aligned carbon nanotubes: a step closer to commercial realization. Chemical Physics Letters 303:467–474.

Bekyarova E, Haddon RC, Parpura V (2005a) Biofunctionalization of Carbon Nanotubes. In: Biofunctionalization of Nanomaterials (Nanotechnologies for the Life Sciences) (Kumar CSSR, ed), pp 41–71. Wienhein-Berlin, Germany: Wiley-VCH.

Bekyarova E, Ni Y, Malarkey EB, Montana V, McWilliams JL, Haddon RC, Parpura V (2005b) Applications of Carbon Nanotubes in Biotechnology and Biomedicine. J Biomed Nanotechnol 1:3–17.

Cellot G, Cilia E, Cipollone S, Rancic V, Sucapane A, Giordani S, Gambazzi L, Markram H, Grandolfo M, Scaini D, Gelain F, Casalis L, Prato M, Giugliano M, Ballerini L (2009) Carbon nanotubes might improve neuronal performance by favouring electrical shortcuts. Nat Nanotechnol 4:126–133.

Fan YW, Cui FZ, Hou SP, Xu QY, Chen LN, Lee IS (2002) Culture of neural cells on silicon wafers with nano-scale surface topograph. J Neurosci Methods 120:17–23.

Galvan-Garcia P, Keefer EW, Yang F, Zhang M, Fang S, Zakhidov AA, Baughman RH, Romero MI (2007) Robust cell migration and neuronal growth on pristine carbon nanotube sheets and yarns. J Biomater Sci Polym Ed 18:1245–1261.

Hu H, Ni Y, Montana V, Haddon RC, Parpura V (2004) Chemically Functionalized Carbon Nanotubes as Substrates for Neuronal Growth. Nano Letters 4:507–511.

Hu H, Ni Y, Mandal SK, Montana V, Zhao B, Haddon RC, Parpura V (2005) Polyethyleneimine Functionalized Single-Walled Carbon Nanotubes as a Substrate for Neuronal Growth. The Journal of Physical Chemistry B 109:4285–4289.

Kater SB, Mattson MP, Cohan C, Connor J (1988) Calcium regulation of the neuronal growth cone. Trends Neurosci 11:315–321.

Lee W, Parpura V (2009) Chapter 6 - Carbon nanotubes as substrates/scaffolds for neural cell growth. Prog Brain Res 180:110–125.

Liu J, Rinzler AG, Dai H, Hafner JH, Bradley RK, Boul PJ, Lu A, Iverson T, Shelimov K, Huffman CB, Rodriguez-Macias F, Shon YS, Lee TR, Colbert DT, Smalley RE (1998) Fullerene pipes. Science 280:1253–1256.

Lovat V, Pantarotto D, Lagostena L, Cacciari B, Grandolfo M, Righi M, Spalluto G, Prato M, Ballerini L (2005) Carbon nanotube substrates boost neuronal electrical signaling. Nano Lett 5:1107–1110.

Lustgarten JH, Proctor M, Haroun RI, Avellino AM, Pindzola AA, Kliot M (1991) Semipermeable polymer tubes provide a microenvironment for in vivo analysis of dorsal root regeneration. J Biomech Eng 113:184–188.

Malarkey EB, Reyes RC, Zhao B, Haddon RC, Parpura V (2008) Water soluble single-walled carbon nanotubes inhibit stimulated endocytosis in neurons. Nano Lett 8:3538–3542.

Malarkey EB, Fisher KA, Bekyarova E, Liu W, Haddon RC, Parpura V (2009) Conductive Single-Walled Carbon Nanotube Substrates Modulate Neuronal Growth. Nano Lett 9:264–268.

Mark RJ, Lovell MA, Markesbery WR, Uchida K, Mattson MP (1997) A role for 4-hydroxynonenal, an aldehydic product of lipid peroxidation, in disruption of ion homeostasis and neuronal death induced by amyloid beta-peptide. J Neurochem 68:255–264.

Matsumoto K, Sato C, Naka Y, Whitby R, Shimizu N (2010) Stimulation of neuronal neurite outgrowth using functionalized carbon nanotubes. Nanotechnology 21:115101.

Matsumoto K, Sato C, Naka Y, Kitazawa A, Whitby RL, Shimizu N (2007) Neurite outgrowths of neurons with neurotrophin-coated carbon nanotubes. J Biosci Bioeng 103:216–220.

Mattson MP, Kater SB (1987) Calcium regulation of neurite elongation and growth cone motility. J Neurosci 7:4034–4043.

Mattson MP, Taylor-Hunter A, Kater SB (1988) Neurite outgrowth in individual neurons of a neuronal population is differentially regulated by calcium and cyclic AMP. J Neurosci 8:1704–1711.

Mattson MP, Haddon RC, Rao AM (2000) Molecular functionalization of carbon nanotubes and use as substrates for neuronal growth. J Mol Neurosci 14:175–182.

Mazzatenta A, Giugliano M, Campidelli S, Gambazzi L, Businaro L, Markram H, Prato M, Ballerini L (2007) Interfacing neurons with carbon nanotubes: electrical signal transfer and synaptic stimulation in cultured brain circuits. J Neurosci 27:6931–6936.

Ni Y, Hu H, Malarkey EB, Zhao B, Montana V, Haddon RC, Parpura V (2005) Chemically functionalized water soluble single-walled carbon nanotubes modulate neurite outgrowth. J Nanosci Nanotechnol 5:1707–1712.

Roman J, Haddon R, Parpura V, Floyd C (2009a) Single Walled Carbon Nanotubes as a Regenerative Substrate in a Rat Model of Spinal Cord Injury. Journal of Neurotrauma 26:A76.

Roman JA, Haddon R, Parpura V, Floyd CL (2009b) Single Walled Carbon Nanotubes as a Regenerative Substrate in Spinal Cord Injury. Journal of Neurochemistry 108:104.

Roman JA, Niedzielko TL, Haddon RC, Parpura V, Floyd CL (2011) Single-Walled Carbon Nanotubes Chemically Functionalized with Polyethylene Glycol Promote Tissue Repair in a Rat Model of Spinal Cord Injury. J Neurotrauma. In Press. doi:10.1089/neu.2010.1409

Sorkin R, Greenbaum A, David-Pur M, Anava S, Ayali A, Ben-Jacob E, Hanein Y (2009) Process entanglement as a neuronal anchorage mechanism to rough surfaces. Nanotechnology 20:8 pp.

Uchida K, Stadtman ER (1992) Modification of histidine residues in proteins by reaction with 4-hydroxynonenal. Proc Natl Acad Sci U S A 89:4544–4548.

Varadan VK, Xie JN (2002) Large-scale synthesis of multi-walled carbon nanotubes by microwave CVD. Smart materials and structures 11:610–616.

Xie JN, Chen LF, Aatre KR, Srivatsan M, Varadan VK (2006) Somatosensory neurons grown on functionalized carbon nanotube mats. Smart materials and structures 15:N85-N88.

Zakharenko S, Popov S (2000) Plasma membrane recycling and flow in growing neurites. Neuroscience 97:185–194.

# Index



G.A. Silva and V. Parpura (eds.), *Nanotechnology for Biology and Medicine: At the Building Block Level*, Fundamental Biomedical Technologies,
DOI 10.1007/978-0-387-31296-5, 

MIX
Papier aus verantwortungsvollen Quellen
Paper from responsible sources
FSC® C105338

If you have any concerns about our products,
you can contact us on
**ProductSafety@springernature.com**

In case Publisher is established outside the EU,
the EU authorized representative is:
**Springer Nature Customer Service Center GmbH**
**Europaplatz 3, 69115 Heidelberg, Germany**

Printed by Libri Plureos GmbH
in Hamburg, Germany